Dimensions of Psychology: Case Studies, Narratives and Experiential Exercises

Dimensions of Psychology: Case Studies, Narratives and Experiential Exercises

D.K. Surana

2008

CYBER TECH PUBLICATIONS
4264/3, Ansari Road, Daryaganj, New Delhi-110002 (India)
Ph.: 011-23244078. Fax: 011-23280028
E-Mail: cyberpublicationsdelhi@yahoo.com

DIMENSIONS OF PSYCHOLOGY : CASE STUDIES, NARRATIVES AND EXPERIENTIAL EXERCISES

D.K. Surana

First Edition 2008

Published by :

G.S. Rawat for **Cyber Tech Publications**
4264/3, Ansari Road, Daryaganj, New Delhi-110002 (India)
Ph.: 011-23244078. Fax: 011-23280028
E-Mail: cyberpublicationsdelhi@yahoo.com

Printed at : **Amit Enterprises**, Maujpur, Delhi, (9810878184)

Preface

Psychology as a discipline specializes in the study of experiences, behaviours and mental processes of human beings within socio-cultural and socio-historical context. This book purports to introduce the learners to the basic ides, principles and methods in psychology so as to enable them to understand themselves and their social world better. The emphasis is put on creating interest and exposure needed by learners to develop their own knowledge base and understanding.

This book deals with psychological knowledge and practices which are contextually rooted. It emphasizes the complexity of behavioural processes and discourages critical reasoning, allowing students to appreciate the role of cultural factors in behaviours, and illustrating how biology and experience shape behaviours.

The present effort at reforming and updating the book and encapsulating all the important topics in a lucid language is based on the feedback received from the teachers, students, as well as some new educational institutions imparting education in psychology.

However, your valuable suggestions for improving the book in its forthcoming edition are always welcome.

—*Author*

Contents

1

Introduction

Psychology is an academic and applied discipline involving the scientific study of mental processes and behavior. Psychology also refers to the application of such knowledge to various spheres of human activity, including problems of individuals' daily lives and the treatment of mental illness.

Psychology differs from the other social sciences—anthropology, economics, political science, and sociology—in seeking to explain the mental processes and behavior of individuals. Psychology differs from biology and neuroscience in that it is primarily concerned with the interaction of mental processes and behavior on a systemic level, as opposed to studying the biological or neural processes themselves. In contrast, the subfield of neuropsychology studies the actual neural processes and how they relate to the mental effects they subjectively produce. Biological psychology is the scientific study of the biological bases of behavior and mental states.

The word psychology comes from the ancient Greek psyche ("soul," "mind") and-ology ("study").

HISTORY

Early Development

The first use of the term "psychology" is often attributed to the German scholastic philosopher Rudolf Goeckel (Latinized Rudolph Goclenius), published in 1590. More than six decades earlier, however, the Croatian humanist Marko Marulic used the term in the title of a work which was subsequently lost. This, of course, may not have been the very first usage, but it is the earliest documented use at present.

The term did not fall into popular usage until the German idealist philosopher, Christian Wolff (1679-1754) used it in his Psychologia empirica and Psychologia rationalis (1732-1734). This distinction between empirical and rational psychology was picked up in Diderot's Encyclopedie and was popularized in France by Maine de Biran. The root of the word psychology (psyche) is very roughly equivalent to "soul" in Greek, and (ology) equivalent to "study". Psychology came to be considered a study of the soul (in a religious sense of this term) much later, in Christian times. Psychology as a medical discipline can be seen in Thomas Willis' reference to psychology (the "Doctrine of the Soul") in terms of brain function, as part of his 1672 anatomical treatise "De Anima Brutorum" ("Two Discourses on the Souls of Brutes"). Until about the end of the 19th century, psychology was regarded as a branch of philosophy.

Early Modern Era: In 1879, Wilhelm Wundt (1832-1920), known as "the father of psychology", founded a laboratory for the study of psychology at Leipzig University in Germany. The American philosopher William James published his seminal book, Principles of Psychology, in 1890, laying the foundations for many of the questions that psychologists would focus on for years to come. Other important early contributors to the field include Hermann Ebbinghaus (1850-1909), a pioneer in the experimental study of memory at the University of Berlin; and the Russian physiologist Ivan Pavlov (1849-1936), who investigated the learning process now referred to as classical conditioning.

Meanwhile, during the 1890s, the Austrian physician Sigmund Freud, who was trained as a neurologist and had no formal training in experimental psychology, had developed a method of psychotherapy known as psychoanalysis. Freud's understanding of the mind was largely based on interpretive methods and introspection, and was focused in particular on resolving mental distress and psychopathology. Freud's theories became very well-known, largely because they tackled subjects such as sexuality and repression as general aspects of psychological development. These were largely considered taboo subjects at the time, and Freud provided a catalyst for them to be openly discussed in polite society. Although Freud's theories are only of limited interest in modern academic psychology departments, his application of psychology to clinical work has been very influential.

Partly in reaction to the subjective and introspective nature of Freudian psychology, and its focus on the recollection of childhood experiences, during the early decades of the 20th century behaviorism gained popularity as a guiding psychological theory. Championed by psychologists such as John B. Watson and Edward Thorndike (and later, B.F. Skinner), behaviorism was grounded in studies of animal behavior. Behaviorists argued that psychology should be a science of behavior, not the mind, and rejected the idea that internal mental states such as beliefs, desires, or goals could be studied scientifically. In his paper "Psychology as the Behaviorist Views It" (1913), Watson argued that psychology "is a purely objective experimental branch of natural science," that "introspection forms no essential part of its methods", and that "the behaviorist recognizes no dividing line between man and brute."

Behaviorism reigned as the dominant model in psychology through the first half of the 20th century, largely due to the creation of conditioning theories as scientific models of human behavior, and their successful application in the workplace and in fields such as advertising.

Modern Era: However, it became increasingly clear that although behaviorism had made some important discoveries, it was deficient as a guiding theory of human behavior. Noam Chomsky's review of Skinner's book Verbal Behavior (that aimed to explain language acquisition in a behaviorist framework) is considered one of the major factors in the ending of behaviorism's reign. Chomsky demonstrated that language could not purely be learned from conditioning, as people could produce sentences unique in structure and meaning that couldn't possibly be generated solely through experience of natural language, implying that there must be internal states of mind that behaviorism rejected as illusory. Similarly, work by Albert Bandura showed that children could learn by social observation, without any change in overt behavior, and so must be accounted for by internal representations.

Humanistic psychology emerged in the 1950s and has continued as a reaction to positivist and scientific approaches to the mind. It stresses a phenomenological view of human experience and seeks to understand human beings and their behavior by conducting qualitative research. The humanistic approach has its roots in existentialist and phenomenological philosophy and many humanist psychologists completely reject a scientific approach, arguing that trying to turn human experience into measurements strips it of all meaning and relevance to lived existence. Some of the founding theorists behind this school of thought were Abraham Maslow who formulated a hierarchy of human needs, Carl Rogers who created and developed Client-centred therapy, and Fritz Perls who helped create and develop Gestalt therapy.

The rise of computer technology also promoted the metaphor of mental function as information processing. This, combined with a scientific approach to studying the mind, as well as a belief in internal mental states, led to the rise of cognitivism as the dominant model of the mind.

Links between brain and nervous system function were also becoming common, partly due to the experimental work of people such as Charles Sherrington and Donald Hebb, and partly due to studies of people with brain injury. With the development of technologies for accurately measuring brain function, neuropsychology and cognitive neuroscience have become some of the most active areas in contemporary psychology. With the increasing involvement of other disciplines (such as philosophy, computer science and neuroscience) in the quest to understand the mind, the umbrella discipline of cognitive science has been created as a means of focusing such efforts in a constructive way.

PRINCIPLES

Mind and Brain: Psychology describes and attempts to explain consciousness, behavior and social interaction. Empirical psychology is primarily devoted to describing human experience and behavior as it actually occurs. In the past 20 years or so psychology has begun to examine the relationship between consciousness and the brain or nervous system. It is still not clear in what ways these interact: does consciousness determine brain states or do brain states determine consciousness-or are both going on in various ways? Perhaps to understand this you need to know the definition of "consciousness" and "brain state"- or is consciousness some sort of complicated 'illusion' which bears no direct relationship to neural processes? An understanding of brain function is increasingly being included in psychological theory and practice, particularly in areas such as artificial intelligence, neuropsychology, and cognitive neuroscience.

Schools of Thought: Various schools of thought have argued for a particular model to be used as a guiding theory by which all, or the majority, of human behavior can be explained. The popularity of these has waxed and waned over time. Some psychologists may think of themselves as adherents to a particular school of thought and reject the others, although most consider each as an approach to understanding the mind, and not necessarily as mutually exclusive theories.

Scope of Psychology: Psychology is an extremely broad field, encompassing many different approaches to the study of mental processes and behavior. Below are the major areas of inquiry that comprise psychology, divided into fields of research psychology and fields of applied psychology. A comprehensive list of the sub-fields and areas within psychology can be found at the list of psychological topics and list of psychology disciplines.

Fields of Research Psychology: Research psychology encompasses the study of behavior for use in academic settings, and contains numerous areas. It contains the areas of abnormal psychology, biological psychology, cognitive psychology, comparative psychology, developmental psychology, personality psychology, social psychology and others. Research psychology is contrasted with applied psychology.

Abnormal Psychology: Abnormal psychology is the study of abnormal behavior in order to describe, predict, explain, and change abnormal patterns of functioning. Abnormal psychology studies the nature of psychopathology and its causes, and this knowledge is applied in clinical psychology to treating patients with psychological disorders.

Biological Psychology: Biological psychology is the scientific study of the biological bases of behavior and mental states. Because all behavior is controlled by the central nervous system, it is sensible to study how the brain functions in order to understand behavior. This is the approach taken in behavioral neuroscience, cognitive neuroscience, and neuropsychology. Neuropsychology is the branch of psychology that aims to understand how the structure and function of the brain relate to specific behavioral and psychological processes. Often neuropsychologists are employed as scientists to advance scientific or medical knowledge. Neuropsychology is particularly concerned with the understanding of brain injury in an attempt to work out normal psychological function. The approach of cognitive neuroscience to studying the link between brain and behavior is to use neuroimaging tools, such as fMRI, to observe which areas of the brain are active during a particular task.

Cognitive Psychology: The nature of thought is another core interest in psychology. Cognitive psychology studies cognition, the mental processes underlying behavior. It uses information processing as a framework for understanding the mind. Perception, learning, problem solving, memory, attention, language and emotion are all well researched areas. Cognitive psychology is associated with a school of thought known as cognitivism, whose adherents argue for an information processing model of mental function, informed by positivism and experimental psychology. Cognitive science is very closely related to cognitive psychology, but differs in some of the research methods used, and has a slightly greater emphasis on explaining mental phenomena in terms of both behavior and neural processing.

Both areas can use computational models to simulate phenomena of interest. Because mental events cannot directly be observed, computational models provide a tool for studying the functional organization of the mind. Such models give cognitive psychologists a way to study the "software" of mental processes independent of the "hardware" it runs on, be it the brain or a computer.

Comparative Psychology: Comparative psychology refers to the study of the behavior and mental life of animals other than human beings. It is related to disciplines outside of psychology that study animal behavior, such as ethology.

Although the field of psychology is primarily concerned with humans, the behavior and mental processes of animals is also an important part of psychological research, either as a subject in its own right (e.g., animal cognition and ethology), or with strong emphasis about evolutionary links, and somewhat more controversially, as a way of gaining an insight into human psychology by means of comparison or via animal models of emotional and behavior systems as seen in neuroscience of psychology (e.g., affective neuroscience and social neuroscience).

Developmental Psychology: Mainly focusing on the development of the human mind through the life span, developmental psychology seeks to understand how people come to perceive, understand, and act within the world and how these processes change as they age. This may focus on intellectual, cognitive, neural, social, or moral development. Researchers who study children use a number of unique research methods to make observations in natural settings or to engage them in experimental tasks. Such tasks often resemble specially designed games and activities that are both enjoyable for the child and scientifically useful, and researchers have even devised clever methods to study the mental processes of small infants. In addition to studying children, developmental psychologists also study aging and processes throughout the life span, especially at other times of rapid change (such as adolescence and old age). Urie Bronfenbrenner's theory of development in context is influential in this field, as are those mentioned in "Educational psychology" immediately below, as well as many others. Developmental psychologists draw on the full range of theorists in scientific psychology to inform their research.

Personality Psychology: Personality psychology studies enduring psychological patterns of behavior, thought and emotion, commonly called an individual's personality. Theories of personality vary between different psychological schools. Trait theories attempts to break personality down into a number of traits, by use of factor analysis. The number of traits have varied between theories. One of the first, and smallest, models was that of Hans Eysenck, which had three dimensions: extraversion-introversion, neuroticism-emotional stability, and psychoticism. Raymond Cattell proposed a theory of 16 personality factors. The theory that has most empirical evidence behind it today may be the "Big Five" theory, proposed by Lewis Goldberg and others.

A different, but well known, approach to personality is that of Sigmund Freud, whose structural theory of personality divided personality into the ego, superego, and id. Freud's theory of personality has been criticized by many, including many mainstream psychologists.

Quantitative Psychology: Quantitative psychology involves the application of statistical analysis to psychological research, and the development of novel statistical approaches for measuring and explaining human behavior. It is a young field (only recently have Ph.D. programs in quantitative psychology been formed), and it is loosely comprised of the subfields psychometrics and mathematical psychology.

Psychometrics is the field of psychology concerned with the theory and technique of psychological measurement, which includes the measurement of knowledge, abilities, attitudes, and personality traits. Measurement of these unobservable phenomena is difficult, and much of the research and accumulated knowledge in this discipline has been developed in an attempt to properly define and quantify such phenomena. Psychometric research typically involves two major research tasks, namely: (i) the construction of instruments and procedures for measurement; and (ii) the development and refinement of theoretical approaches to measurement.

Social Psychology: Social psychology is the study of the nature and causes of human social behavior, with an emphasis on how people think towards each other and how they relate to each other. Social Psychology aims to understand how we make sense of social situations. For example, this could involve the influence of others on an individual's behavior (e.g., conformity or persuasion), the perception and understanding of social cues, or the formation of attitudes or stereotypes about other people. Social cognition is a common approach and involves a mostly cognitive and scientific approach to understanding social behavior.

A related area is community psychology, which examines psychological and mental health issues on the level of the community rather than using the individual as the unit of measurement. "Sense of community" has become its conceptual center (Sarason, 1986; Chavis & Pretty, 1999).

Fields of Applied Psychology: Applied psychology encompasses both psychological research that is designed to help individuals overcome practical problems and the application of this research in applied settings. Much of applied psychology research is utilized in other fields, such as business management, product design, ergonomics, nutrition, and clinical medicine. Applied psychology includes the areas of clinical psychology, industrial and organizational psychology, human factors, forensic psychology, health psychology, school psychology and others.

Clinical Psychology: Clinical psychology is the application of abnormal psychology research to the understanding, treatment, and assessment of psychopathology, including behavioral and mental health concerns. It has traditionally been associated with psychological treatment and psychotherapy, although modern clinical psychology may take an eclectic approach, including a number of therapeutic approaches. Typically, although working with many of the same clients as psychiatrists, clinical psychologists do not prescribe psychiatric drugs. Some clinical psychologists may focus on the clinical management of patients with brain injury. This area is known as clinical neuropsychology.

In recent years and particularly in the United States, a major split has been developing between academic research psychologists in universities and some branches of clinical psychology. Many research psychologists believe that many contemporary clinicians use therapies based on discredited theories and unsupported by empirical evidence of their effectiveness. From the other side, these clinicians believe that the research psychologists are ignoring their experience in dealing with actual patients. The disagreement resulted in the formation of the Association for Psychological Science by the research psychologists as a new body distinct from the American Psychological Association.

The work performed by clinical psychologists tends to be done inside various therapy models. A popular model is the

Cognitive-Behaviorial therapy (CBT) framework. CBT is an umbrella term that refers to a number of therapies which focus on changing cognitions and/or behaviors, rather than changing behavior exclusively, or discovering the unconscious causes of psychopathology (as in the psychodynamic school). The two most famous CBT therapies are Aaron T. Beck's cognitive therapy and Albert Ellis's rational emotive behavior therapy (with cognitive therapy being, by far, the most extensively studied therapy in contemporary clinical psychology).

Counseling Psychology: Counseling psychology is a psychology specialty that facilitates personal and interpersonal functioning across the lifespan with a focus on emotional, social, vocational, educational, health-related, developmental, and organizational concerns. Counseling psychology differs from clinical psychology in that it is focused more on normal developmental issues and everyday stress rather than serious mental disorders. Counseling psychologists are employed in a variety of settings, including universities, schools, businesses, private practice, and community mental health centers.

The emerging field of relationship counseling, which characterizes ordinary human relationship successes and failures in concrete terms, has the specific appeal of avoiding psychology's practice of ascribing pathology to individuals who seek assistance. Current health insurance reimbursement for psychological services commonly involves the assignment of mental disease nomenclature(a feature that potential clients may find offensive, and that could potentially be iatrogenic).

Relationship counseling, also referred to as "relationship education", includes psychologists, psychiatrists, and social workers. It is based on decades of university-based research, drawing on knowledge gained through close observation and analysis of both successful and unsuccessful marriages and family units.

Psychology and Law: Together, Forensic psychology and Legal Psychology compose the area known as Psychology and Law. Most typically, forensic psychology is practiced by clinical psychologists, and involves a clinical analysis of a particular individual and an assessment of some specific psycho-legal question. Typically, referrals to forensic practices constitute assessments for individuals that have ostensibly suffered neurologic insult(s). These patients have sought legal recourse, and the job of the forensic psychologist is to demonstrate that there is or is not (depending on their employment by either the prosecution or defense) a cause-and-effect relation between the accident and the subsequent (again, ostensible) neurologic change. Legal psychology is a research-oriented field populated with researchers from several different areas within psychology (although Social Psychologists are typical).

Health Psychology: Health psychology is the application of psychological theory and research to health, illness and health care. Whereas clinical psychology focuses on mental health and neurological illness, health psychology is concerned with the psychology of a much wider range of health-related behavior including healthy eating, the doctor-patient relationship, a patient's understanding of health information, and beliefs about illness. Health psychologists may be involved in public health campaigns, examining the impact of illness or health policy on quality of life or in research into the psychological impact of health and social care.

Human Factors Psychology: Human factors psychology is the study of how cognitive and psychological processes affect our interaction with tools and objects in the environment. The goal of research in human factors psychology is to better design objects by taking into account the limitations and biases of human mental processes and behavior.

Industrial and Organizational Psychology: Industrial and organizational psychology (I/O) is among the newest fields in psychology. Industrial Psychology focuses on improving, evaluating, and predicting job performance while Organizational Psychology focuses on how organizations impact and interact with individuals. In 1910, through the works and experiments of Hugo Munsterberg and Walter Dill Scott, Industrial Psychology became recognized as a legitimate part of the social science. Organizational Psychology was not officially added until the 1970s and since then, the field has flourished. The Society for Industrial Organizational Psychology has approximately 3400 professional members and 1900 student members. These two numbers combine to make up only about four percent of the members in the American Psychology Association but the number has been rising since 1939 when there were only one hundred professional I/O psychologists.

I/O psychologists are employeed by academic institutions, consulting firms, internal human resources in industries, and governmental institutions. Various universities across the United States are beginning to strengthen their I/O Psychology programs due to the increase of interest and job demand in the field. Industrial organizational psychologists look at questions regarding things such as who to hire, how to define and measure successful job performance, how to prepare people to be more successful in their jobs, how to create and change jobs so that they are safer and make people happier, and how to structure the organization to allow people to achieve their potential.

School Psychology: School psychology is the area of discipline that is dedicated to helping young people succeed academically, socially, and emotionally. School psychologists collaborate with educators, parents, and other professionals to create safe, healthy, and supportive learning environments for all students that strengthen connections between home and school. They are trained to be experts in educational and behavioral assessment, intervention, prevention, and consultation, and many have extensive training in research. Currently, school psychology is the only field in which a professional can be called a "psychologist" without a doctoral degree, with the National Association of School Psychologists (NASP) recognizing the Specialist degree as the entry level. This is a matter of controversy, however, as the APA does not recognize anything below a doctorate as the entry level for a psychologist. Specialist-level school psychologists, who typically receive three years of graduate training, function almost exclusively within school systems, while those at the doctoral-level are found in a number of other settings as well, including universities, hospitals, clinics, and private practice.

RESEARCH METHODS OF PSYCHOLOGY

Research in psychology is conducted in broad accord with the standards of the scientific method, encompassing both qualitative ethological and quantitative statistical modalities to generate and evaluate explanatory hypotheses with regard to psychological phenomena. Where research ethics and the state of development in a given research domain permits, investigation may be pursued by experimental protocols. Psychology tends to be eclectic, drawing on scientific knowledge from other fields to help explain and understand psychological phenomena. Qualitative psychological research utilizes a broad spectrum of observational methods, including action research, ethography, exploratory statistics, structured interviews, and participant observation, to enable the gathering of rich information unattainable by classical experimentation. Research in humanistic psychology is more typically pursued by ethnographic, historical, and historiographic methods. The testing of different aspects of psychological function is a significant area of contemporary psychology. Psychometric and statistical methods predominate, including various well-known standardized tests as well as those created ad hoc as the situation or experiment requires.

Academic psychologists may focus purely on research and psychological theory, aiming to further psychological understanding in a particular area, while other psychologists may work in applied psychology to deploy such knowledge for immediate and practical benefit. However, these approaches are not mutually exclusive and most psychologists will be involved in both researching and applying psychology at some point during their career. Clinical psychology, among many of the various disciplines of psychology, aims at developing in practicing psychologists knowledge of and experience with research and experimental methods which they will continue to build up as well as employ as they treat individuals with psychological issues or use psychology to help others.

When an area of interest requires specific training and specialist knowledge, especially in applied areas, psychological associations normally establish a governing body to manage training requirements. Similarly, requirements may be laid down for university degrees in psychology, so that students acquire an adequate knowledge in a number of areas. Additionally, areas of practical psychology, where psychologists offer treatment to others, may require that psychologists be licensed by government regulatory bodies as well.

Controlled Experiments: Experimental psychological research is conducted in the laboratory under controlled conditions. This method of research attempts to rely solely on an application of the scientific method to understand behavior and mental processes. Examples of such measurements of behavior include reaction time and various psychometric measurements. Experiments are conducted to test a particular hypothesis.

As an example of a psychological experiment, one may want to test people's perception of different tones. Specifically, one could ask the following question: is it easier for people to discriminate one pair of tones from another depending upon their frequency? To answer this, one would want to disprove the hypothesis that all tones are equally discriminable, regardless of their frequency. A task to test this hypothesis would have a participant seated in a room listening to a series of tones. If the participant would make one indication (by pressing a button, for example) if they thought the tones were two different sounds, and another indication if they thought they were the same sound. The proportion of correct responses would be the measurement used to describe whether or not all the tones were equally discriminable. The result of this particular experiment would probably indicate better discrimination of certain tones based on the human threshold of hearing.

Longitudinal Studies: A longitudinal study is a research method which observes a particular population over time. For example, one might wish to study specific language impairment (SLI) by observing a group of individuals with the condition over a period of time. This method has the advantage of seeing how a condition can affect individuals over long time scales. However, such studies can suffer from attrition due to drop-out or death of subjects. In addition, since individual differences between members of the group are not controlled, it may be difficult to draw conclusions about the populations.

Neuropsychological Methods: Neuropsychology involves the study of both healthy individuals and patients, typically who have suffered either brain injury or mental illness.

Cognitive neuropsychology and cognitive neuropsychiatry study neurological or mental impairment in an attempt to infer theories of normal mind and brain function. This typically involves looking for differences in patterns of remaining ability (known as 'functional disassociation's') which can give clues as to whether abilities are comprised of smaller functions, or are controlled by a single cognitive mechanism.

In addition, experimental techniques are often used which also apply to studying the neuropsychology of healthy individuals. These include behavioral experiments, brain-scanning or functional neuroimaging-used to examine the activity of the brain during task performance, and techniques such as transcranial magnetic stimulation, which can safely alter the function of small brain areas to investigate their importance in mental operations.

Computational Modeling: Computational modeling is a tool often used in cognitive psychology to simulate a particular behavior using a computer. This method has several advantages. Since modern computers are extremely fast, many simulations can be run in a short time, allowing for a great deal of statistical power. Modeling also allows psychologists to visualise hypotheses about the functional organization of mental events that couldn't be directly observed in a human.

Several different types of modeling are used to study behavior. Connectionism uses neural nets to simulate the brain. Another method is symbolic modeling, which represents many different mental objects using variables and rules. Other types of modeling include dynamic systems and stochastic modeling.

CRITICISM

Controversy as a Science: Although modern psychology attempts to be a scientific endeavour, the field has a history of controversy. Some criticisms of psychology have been made on ethical and philosophical grounds. Some have argued that by subjecting the human mind to experimentation and statistical study, psychologists objectify persons. Because it treats human beings as things, as objects that can be examined by experiment, psychology is sometimes portrayed as dehumanizing, ignoring or downplaying what is most essential about being human. This criticism has come from within the field as well, particularly by existential and humanistic psychologists.

A common criticism of psychology concerns its fuzziness as a science. Philosopher Thomas Kuhn suggested in 1962 that psychology is in a pre-paradigmatic state, lacking the agreement on facts found in mature sciences such as chemistry and physics. Because some areas of psychology rely on "soft" research methods such as surveys and questionnaires, critics have claimed that psychology is not as scientific as psychologists assume. Methods such as introspection and psychoanalysis, used by some psychologists, are inherently subjective. Objectivity, validity, and rigour are key attributes in science, and some approaches to psychology have fallen short on these criteria. On the other hand, greater use of statistical controls and increasingly sophisticated research design, analysis, and statistical methods, as well as a decline (at least within academic psychology departments) in the use of less scientific methods, have lessened the impact of this criticism to some degree.

Debates continue, however, such as the questioned effectiveness of probability testing as a valid research tool. The concern is that this statistical method may promote trivial findings as meaningful, especially when large samples are used. Psychologists have responded with an increased use of effect size statistics, rather than sole reliance on the traditional $p<.05$ decision rule.

Research-practitioner Gap: There is also concern from researchers about a perceived scientific gap between empirically based practices. Exponents of evidence-based approaches to psychological practice say that "over the past several decades, the fields of clinical psychology, psychiatry, and social work have borne witness to a widening and deeply troubling gap between science and practice" and "less and less of what researchers do finds its way into the consulting room, and less and less of what practitioners do derives from scientific evidence." Moreover there are many "unvalidated and sometimes harmful psychotherapeutic methods" that have been widely adopted by the profession. However, "the fields of clinical psychology, psychiatry, and social work have recently placed increased emphasis on evidence-based mental health practices."

2

Psychological Thoughts

The history of psychology as a scholarly study of the mind and behavior dates, in Europe, back to the Late Middle Ages. It was widely regarded to a branch of philosophy until the middle of the 19th-century when a scientific and eventually experimental form of the discipline emerged in Germany. Psychology borders on various other fields including physiology, neuroscience, artificial intelligence, sociology, and anthropology.

Overview: During the last quarter of the 19th century, psychology in the West began to be seriously pursued as a scientific enterprise. Psychology as an experimental field of study is commonly seen as starting in 1879, when Wilhelm Wundt founded the first laboratory dedicated exclusively to psychological research in Leipzig. Other important early contributors to the field include Hermann Ebbinghaus (a pioneer in studies on memory), William James, and Ivan Pavlov (who developed the procedures associated with classical conditioning). Sigmund Freud's psychoanalysis, though widely known, has had a contested relationship with the development of psychology.

Soon after the development of experimental psychology, various kinds of applied psychology began to appear as well. G. Stanley Hall (Johns Hopkins) brought scientific pedagogy to the United States from Germany in the early 1880s. John Dewey's (U. Chicago) educational theory of the 1890s was an early example as well. Also in the 1890s, Hugo Münsterberg (Harvard) began writing about the application of psychology to industry, law, and other professions. Lightner Witmer (U. Pennsylvania) established the first psychological clinic in 1900. James McKeen Cattell (Columbia U.) adapted the Francis

Galton's anthropometric methods to generate the first program of mental testing in the 1890s as well.

The 20th century saw a rejection of Freud's theories as being too unscientific, and a reaction against Edward Titchener's abstract approach to the mind. This led to the formulation of behaviorism by John B. Watson, which was popularized by B.F. Skinner. Behaviorism proposed epistemologically limiting psychological study to overt behavior, since that could be quantified and easily measured. Scientific knowledge of the "mind" was considered too metaphysical, hence impossible to achieve. The final decades of the 20th century have seen the rise of a new interdisciplinary approach to studying human psychology, known collectively as cognitive science. Cognitive science again considers the "mind" as a subject for investigation, using the tools of evolutionary psychology, linguistics, computer science, philosophy, and neurobiology. This new form of investigation has proposed that a wide understanding of the human mind is possible, and that such an understanding may be applied to other research domains, such as artificial intelligence.

EARLY PSYCHOLOGICAL THOUGHT

Many cultures throughout history have speculated on the nature of the mind, soul, spirit, etc. For instance, in Ancient Egypt, the Edwin Smith Papyrus contains an early description of the brain, and some speculations on its functions (though in a medical/surgical context). The Ebers papyrus (ca 1550 BC) contains a short description of clinical depression. Though full of incantations and applications meant to turn away disease-causing demons and other superstition, it also shows a long tradition of empirical practice and observation.

Ancient Greek philosophers from Thales (fl. 550 bc) through to the Roman period developed an elaborate theory of what they termed the psuch? (from which the first half of "psychology" is derived), as well as other (loosely speaking) "psychological" terms--nous, thumos, logistikon, etc. The most influential of these are the accounts of Plato and of Aristotle. Hellenistic philosophers (viz., the Stoics and Epicurians) diverged from the Classical Greek tradition in several important ways, esp. in being more concerned with questions of the physiological basis of the mind. The Roman physician Galen addressed these issues most elaborately and influentially of all. The Greek tradition influenced some Christian and Islamic thought on the topic.

The Dead Sea Scrolls Manual of Discipline (ca. 100 BC) notes the division of human nature into two temperaments.

In Asia, China had a long history of administering psychological tests as part of its education system. In the 6th century AD, Lin Xie carried out an early psychological experiment, in which he asked people to draw a square with one hand and at the same time draw a circle with the other in order to test people's vulnerability to distraction.

India, too, had an elaborate theory of the self in its Vedanta philosophical writings.

THE BEGINNINGS OF WESTERN PSYCHOLOGY

The scholarly discipline known specifically as "psychology," however, is a European invention. The first use of the term "psychology" is often attributed to the "Yucologia hoc est de hominis perfectione, anima, ortu", written by the German scholastic philosopher Rudolf Göckel (1547-1628, often known under the Latin form Rudolph Goclenius), and published in Marburg in 1590. However, the term was in fact used more than six decades earlier by the Croatian humanist Marko Marulic (1450-1524) in the title of his Latin treatise "Psichiologia de ratione animae humanae." Although the treatise itself has not been preserved, its title appears in a list of Marulic's works compiled by his younger contemporary, Franjo Bozicevic-Natalis in his "Vita Marci Maruli Spalatensis" (Krstic, 1964). This, of course may well not have been the very first usage, but it is the earliest documented use at present.

The term did not come into popular usage until the German idealist philosopher, Christian Wolff (1679-1754) used it in his Psychologia empirica and Psychologia rationalis (1732-1734). This distinction between empirical and rational psychology was picked up in Denis Diderot's (1713-1780) Encyclopédie (1751-1784) and was popularized in France by Maine de Biran (1766-1824). In England, the term "psychology" did not overtake "mental philosophy" until the middle of the 19th century, esp. in the work of William Hamilton (1788-1856).

Early psychology was regarded as the study of the soul (in the Christian sense of the term). The emergence of the modern philosophical form of psychology can be said to have started with the works of René Descartes (1596-1650), of which the most relevant were Meditations on First Philosophy (1641), Passions of the Soul (1649) and Treatise on Man (completed in 1632 but, along with the rest of The World, withheld from publication after Descartes heard of the Catholic Church's condemnation of Galileo).

The emergence of psychology as a medical discipline can be seen in Thomas Willis' reference to psychology (the "Doctrine of the Soul") in terms of brain function, as part of his 1672 anatomical treatise "De Anima Brutorum" ("Two Discourses on the Souls of Brutes").

The philosophers of the British Empiricist and Associationist schools had a profound impact on the later course of experimental psychology. John Locke's An Essay Concerning Human Understanding (1689), George Berkeley's Treatise Concerning the Principles of Human Knowledge (1710), and David Hume's A Treatise of Human Nature (1739-1740) were particularly influential, as were David Hartley's Observations on Man (1749) and John Stuart Mill's A System of Logic. (1843). Also notable was the work of some Continental Rationalist philosophers, especially Baruch Spinoza's (1632-1677) On the Improvement of the Understanding (1662) and Gottfried Wilhelm Leibniz's (1646-1716) New Essays on Human Understanding (completed 1705, published 1765).

Also influential on the emerging discipline of psychology were debates surrounding the efficacy of Mesmerism (hypnosis) and the value of phrenology. The former was developed in the 1770s by Austrian physician Anton Mesmer (1734-1815) who claimed to use the power of gravity, and later of "animal magnetism," to cure various physical and mental ills. As Mesmer and his treatment became increasingly fashionable in both Vienna and Paris, it also began to come under the scrutiny of suspicious officials. In 1784, an investigation was commissioned in Paris by King Louis which included American ambassador

Benjamin Franklin, chemist Antoine Lavoisier and physician Joseph-Ignace Guillotin (later the popularizer of the guillotine). They concluded that Mesmer's method was useless. Although discredited, the "magnetic" tradition continued among Mesmer's students and others, resurfacing in England in the 19th century in the work of physicians John Elliotson (1791-1868), James Esdaile (1808-1859), and James Braid (1795-1860), who renamed it "hypnotism." Mesmerism also continued to have a strong social (if not medical) following in England through the 19th century. In France the practice regained a strong following after it was investigated by the Nancy physician Hippolyte Bernheim (1840-1919) and adopted for the treatment of hysteria by the director of Paris's Salpêtrière Hospital, Jean-Martin Charcot (1825-1893).

Phrenology began as "organology", a theory of brain structure developed by the German physician, Franz Joseph Gall (1758-1828). Gall argued that the brain is divided into a large number of functional "organs," each responsible for particular human mental abilities and dispositions--hope, love, spirituality, greed, language, the abilities to detect the size, form, and color of objects, etc. He argued that the larger each of these organs are, the greater the power of the corresponding mental trait. Further, he argued that one could detect the sizes of the organs in a given individual by feeling the surface of that person's skull. Gall's ultra-localizationist position with respect to the brain was soon attacked, most notably by French anatomist Pierre Flourens (1794-1867), who conducted ablation studies (on chickens) which purported to demonstrate little or no cerebral localization of function. Although Gall had been a serious (if misguided) researcher, his theory was taken by his assistant, Johann Gaspar Spurzheim (1776-1832), and developed into the profitable, popular enterprise of phrenology, which soon spawned, especially in Britain, a thrinving industry of independent practitioners. In the hands of Socttish religious leader George Combe (1788-1858) (whose book The Constitution of Man was one of the best-sellers of the century), phrenology became strongly associated with political reform movements and egalitarian principles. Phrenology soon spread to America as well, where itinerant practical phrenologists assessed the mental well-being of willing customers.

THE EMERGENCE OF GERMAN EXPERIMENTAL PSYCHOLOGY

Until the middle of the 19th century, psychology was widely regarded as a branch of philosophy. Immanuel Kant (1724-1804), for instance, famously declared in his Metaphysical Foundations of Natural Science (1786) that a scientific psychology "properly speaking" (a phrase in which much is buried) is an impossibility. He then proceeded to produce what looks to modern eyes very much like an empirical (if not "properly" scientific) psychology in his Anthropology.

Johann Friedrich Herbart (1776-1841) took issue with Kant's conclusion and attempted to develop a mathematical basis for a scientific psychology. Although he was unable to render his theory empirically testable, his efforts influenced later German scientists. Ernst Heinrich Weber (1795-1878) and Gustav Theodor Fechner (1801-1887), for instance, attempted to measure the mathematical relationships between the physical magnitudes of external stimuli and the psychological intensities of the resulting senstations. Fechner (1860) named this endeavor psychophysics.

Meanwhile, individual differences in reaction time had become a critical issue in the field of astronomy, under the name of the "personal equation". Early researches by Friedrich Wilhelm Bessel (1784-1846) in Königsberg and Adolf Hirsch led to the invention of highly precise chronoscope by Mathias Hipp (based on a design by Charles Wheatstone for a device that measured the speed of artillery shells). Other timing instruments were borrowed from physiology (e.g., the kymograph) and adapted for use by the Utrecht ophthamologist Franciscus Cornelis Donders (1818-1899) and his student Johan Jacob de Jaager in measuring the duration of simple mental decisions.

The 19th century was also the period in which physiology, including neurophysiology, professionalized and saw some of its most significant discoveries. Among its leaders were Charles Bell (1774-1843) and François Magendie (1783-1855) who independently discovered the distinction between sensory and motor nerves in the spinal column, Johannes Müller (1801-1855) who proposed the doctrine of specific nerve energies, Emil du Bois-Reymond (1818-1896) who studied the electrical basis of muscle contraction, Pierre Paul Broca (1824-1880) and Carl Wernicke (1848-1905) who identified areas of the brain responsible for different aspects of language, as well as Gustav Frisch (1837-1927), Eduard Hitzig (1839-1907), and David Ferrier (1843-1924) who localized sensory and motor areas of the brain. One of the principal founders of experimental physiology, Hermann von Helmholtz (1821-1894), conducted studies of a wide range of topics that would later be of interest to psychologists--the speed of neural transmission, the natures of sound and color, and of our perceptions of them, etc. In the 1860s, while he held a position in Heidelberg, Helmholtz engaged as an assistant a young M.D. named Wilhelm Wundt. Wundt employed the equipment of the physiology laboratory--chronoscope, kymograph, and various peripheral devices--to address more complicated psychological questions than had heretofore been considered experimentally. In particular he was interested in the nature of apperception--the point at which a perception comes into the central focus of conscious awareness.

In 1874 Wundt took up a professorship in Zurich, where he published his landmark textbook, Grundzüge der physiologischen Psychologie (Principles of Physiological Psychology, 1874). Moving to a more prestigious professorship in Leipzig in 1875, Wundt founded a laboratory specifically dedicated to original research in experimental psychology in 1879, the first laboratory of its kind in the world. In 1883, he launched a journal in which to publish the results of his, and his students', research, Philosophische Studien (Philosophical Studies). Wundt attracted a large number of students not only from Germany, but also from abroad. Among his most influential American students were Granville Stanley Hall (who had already obtained a PhD from Harvard under the supervision of William James), James McKeen Cattell (who was Wundt's first assistant), and Frank Angell. The most influential British student was Edward Bradford Titchener (who later became professor at Cornell). Experimental psychology laboratories were soon also established at Berlin by Carl Stumpf (1848-1936) and at Göttingen by Georg Elias Müller (1850-1934). Another major German experimental psychologist of the era, though he did not direct his own research institute, was Hermann Ebbinghaus (1850-1909). Experimentation was by no means the only approach to psychology in the German-speaking world at this time. Starting in the 1890s, employing the traditional medical method of the case study, the Viennese physician-Sigmund Freud developed and applied a method of uncovering repressed wishes known as psychoanalysis. Freud's understanding of the mind was largely based on interpretive methods, but was particularly focused on treatment of individuals' psychological problems. Freud's theories were notable for their emphasis on the roles of the individual's unconscious and sexuality. While Freud's work remains scientifically controversial-with many modern-day psychologists and philosophers of science seeing it as being unscientific (being arguably unfalsifiable)-there is no question of the huge and lasting cultural influence it has had.

EARLY AMERICAN PSYCHOLOGY

Around 1875, the Harvard physiology instructor (as he then was), William James, opened a small experimental psychology demonstration laboratory for use with his courses. The laboratory was never used, in those days, for original research, and so controversy remains as to whether it is to be regarded as the "first" experimental psychology laboratory or not. In 1878, James gave a series of lectures at Johns Hopkins University entitled "The Senses and the Brain and their Relation to Thought" in which he argued, contra Thomas Henry Huxley, that consciousness is not epiphenomenal, but must have an evolutionary function, or it would not have been naturally selected in humans. The same year James was contracted by Henry Holt to write a textbook on the "new" experimental psychology. If he had written it quickly, it would have been the first English-language textbook on the topic. It was twelve years, however, before his two-volume Principles of Psychology would be published. In the meantime textbooks were published by George Trumbull Ladd of Yale (1887) and James Mark Baldwin then of Lake Forest College (1889).

In 1879 Charles Sanders Peirce was hired as a philosophy instructor at Johns Hopkins University. Although better known for his astronomical and philosophical work, Peirce also conducted what are perhaps the first American psychology experiments, on the subject of color vision, published in 1877 in the American Journal of Science. Peirce and his student Joseph Jastrow published "On Small Differences in Sensation" in the Memoirs of the National Academy of Sciences, in 1884. In 1882, Peirce was joined at Johns Hopkins by Granville Stanley Hall, who opened the first American research laboratory devoted to experimental psychology in 1883. Peirce was forced out of his position by scandal and Hall was awarded the only professorship in philosophy at Johns Hopkins. In 1887 he founded the American Journal of Psychology, which published work primarily emanating from his own laboratory. In 1888 he left Johns Hopkins for the presidency of the newly-founded Clark University, where he remained for the rest of his career.

Soon, experimental psychology laboratories were opened at the University of Pennsylvania (in 1887, by James McKeen Cattell), Indiana University (1888, William Lowe Bryan), the University of Wisconsin (1888, Joseph Jastrow), Clark University (1889, Edmund Clark Sanford), the McLean Asylum (1889, William Noyes), and the University of Nebraska (1889, Harry Kirk Wolfe).

In 1890, William James' Principles of Psychology finally appeared, and rapidly became the most influential textbook in the history of American psychology. It laid many of the foundations for the sorts of questions that American psychologists would focus on for years to come. The book's chapters on consciousness, emotion, and habit were particularly agenda-setting.

One of those who felt the impact of James' Principles was John Dewey, then professor of philosophy at the University of Michigan. With his junior colleagues, James Hayden Tufts (who founded the psychology laboratory at Michigan) and George Herbert Mead, and his student James Rowland Angell, this group began to reformulate psychology, focusing more strongly on the social environment and on the activity of mind and behavior than the psychophysics-inspired physiological psychology of Wundt and his followers had heretofore. Tufts left Michigan for another junior position at the newly-founded University of Chicago in 1892. A year later, the senior philosopher at Chicago resigned, and Tufts recommended to Chicago president William Rainey Harper that Dewey be offered the position. After initial reluctance, Dewey was hired in 1894. Dewey soon filled out the department with his Michigan companions Mead and Angell. These four formed the core of the Chicago School of psychology.

In 1892, G. Stanley Hall invited 30-some psychologists and philosophers to a meeting at Clark with the purpose of founding a new American Psychological Association (APA). The first annual meeting of the APA was held later that year, hosted by George S. Fullerton at the University of Pennsylvania. Almost immediately tension arose between the experimentally-and philosophically-inclined members of the APA. Edward Bradford Titchener and Lightner Witmer launched an attempt to either establish a separate "Section" for philosophical presentations, or to eject the philosophers altogether. After nearly a decade of debate a Western Philosophical Association was founded and held its first meeting in 1901 at the University of Nebraska. The following year (1902), an American Philosophical Association held its first meeting at Columbia University. These ultimately became the Central and Eastern Divisions of the modern American Philosophical Association.

In 1894, a number of psychologists, unhappy with the parochial editorial policies of the American Journal of Psychology approached Hall about appointing an editorial board and opening the journal out to more psychologists not within Hall's immediate circle. Hall refused, so James McKeen Cattell (then of Columbia) and James Mark Baldwin (then of Princeton) co-founded a new journal, Psychological Review, which rapidly grew to become a major outlet for American psychological researchers.

Beginning in 1895, James Mark Baldwin (Princeton) and Edward Bradford Titchener (Cornell) entered into an increasingly acrimonious dispute over the correct interpretation of some anomalous reaction time findings that had come from the Wundt laboratory (originally reported by Ludwig Lange and James McKeen Cattell). In 1896, James Rowland Angell and Addison W. Moore (Chicago) published a series of experiments in Psychological Review appearing to show that Baldwin was the more correct of the two. However, they interpreted their findings in light of John Dewey's new approach to psychology, which rejected the traditional stimulus-response understanding of the reflex arc in favor of a "circular" account in which what serves as "stimulus" and what as "response" depends on how one views the situation. The full position was laid out in Dewey's landmark article "The Reflex Arc Concept in Psychology" which also appeared in Psychological Review in 1896.

Titchener responded in Philosophical Review (1898, 1899) by distinguishing his austere "structural" approach to psychology from what he termed the Chicago group's more applied "functional" approach, and thus began the first major theoretical rift in American psychology between Structuralism and Functionalism. The group at Columbia, led by James McKeen Cattell, Edward L. Thorndike, and Robert S. Woodworth, was often regarded as a second (after Chicago) "school" of American Functionalism, although they never used that term themselves, because their research focused on the applied areas of mental testing, learning, and education. Dewey was elected president of the APA in 1899, while Titchener dropped his membership in the association. (In 1904, Titchener formed his own group, eventually known as the Society of Experimental Psychologists.) Jastrow promoted the functionalist approach in his APA presidential address of 1900, and Angell adopted Titchener's label explicitly in his influential textbook of 1904 and his APA presidential address of 1906. In reality, Structuralism was, more or less, confined to Titchener and his students. Functionalism, broadly speaking, with its more practical emphasis on action and application, better suited the American cultural "style" and, perhaps more important, was more popular among university trustees and private funding agencies.

EARLY FRENCH PSYCHOLOGY

In no small measure because of the conservatism of the reign of Louis Napoléon (president, 1848-1852; emperor as "Napoléon III," 1852-1870), academic philosophy in France through the middle part of the 19th century was controlled by members of the eclectic and spiritualist schools, led by figures such as Victor Cousin (1792-1867), Théodore Jouffroy (1796-1842), and Paul Janet (1823-1899). These were traditional metaphysical schools, opposed to regarding psychology as a natural science. With the ouster of Napoléon III after the débacle of the Franco-Prussian war, new paths, both political and intellectual, became possible. From the 1870 forward, a steadily increasing interest in positivist, materialist, evolutionary, and deterministic approaches to psychology developed, influenced by, among others, the work of Hyppolyte Taine (1828-1893) (e.g., De L'Intelligence, 1870) and Théodule Ribot (1839-1916) (e.g., La Psychologie Anglaise Contemporaine, 1870).

In 1876, Ribot founded Revue Philosophique (the same year as Mind was founded in Britain), which for the next generation would be virtually the only French outlet for the "new" psychology (Plas, 1997). Although not a working experimentalist himself, Ribot's many books were to have profound influence on the next generation of psychologists. These included especially his L'Hérédité Psychologique (1873) and La Psychologie Allemande Contemporaine (1879). In the 1880s, Ribot's interests turned to psychopathology, writing books on disorders of memory (1881), will (1883), and personality (1885), and where he attempted to bring to these topics the insights of general psychology. Although in 1881 he lost a Sorbonne professorship in the History of Psychological Doctrines to traditionalist Jules Soury (1842-1915), from 1885 to 1889 he taught experimental psychology at the Sorbonne. In 1889 he was awarded a chair at the Collège de France in Experimental and Comparative Psychology, which he held until 1896 (Nicolas, 2002).

France's primary psychological strength lay in the field of psychopathology. The chief neurologist at the Salpêtrière Hospital in Paris, Jean-Martin Charcot (1825-1893), had been using the recently revivied and renamed practice of hypnoisis to "experimentally" produce hysterical symptoms in some of his patients. Two of his students, Alfred Binet (1857-1911) and Pierre Janet (1859-1947), adopted and expanded this practice in their own work.

In 1889, Binet and his colleague Henri Beaunis (1830-1921) co-founded, at the Sorbonne, the first experimental psychology laboratory in France. Just five years later, in 1894, Beaunis,

Binet, and a third colleague, Victor Henri (1872-1940), co-founded the first French journal dedicated to experimental psychology, L'Année Psychologique. In the first years of the 20th century, Binet was requested by the French government to develop a method for the newly-founded universal public education system to identify students who would require extra assistance to master the standardized curriculum. In response, with his collaborator Théodore Simon (1873-1961), he developed the Binet-Simon Intelligence Test, first published in 1905 (revised in 1908 and 1911).

Although the test was used to effect in France, it would find its greatest success (and controversy) in the United States, where it was translated in by Henry H. Goddard (1866-1957), the director of the Training School for the Feebleminded in Vineland, New Jersey, and his assistant, Elizabeth Kite (a translation of the 1905 edition appeared in the Vineland Bulletin in 1908, but much better known was Kite's 1916 translation of the 1908 edition, which appeared in book form). The translated test was used by Goddard to advance his eugenics agenda with respect to those he deemed congenitally unintelligent, especially immigrants from non-Western European countries. Binet's test was revised by Stanford professor Lewis M. Terman (1877-1956) into the Stanford-Binet IQ test in 1916.

With Binet's death in 1911, the Sorbonne laboratory and L'Annee Psychologique fell to Henri Pieron (1881-1964). Piéron's orientation was more physiological that Binet's had been.

Pierre Janet became the leading psychiatrist in France, being appointed to the Salpêtrière (1890-1984), the Sorbonne (1895-1920), and the Collège de France (1920-1936). In 1904, he co-founded the Journale de Psychologie Normale et Pathologique with fellow Sorbonne professor Georges Dumas (1866-1946), a student and faithful follower of Ribot. Whereas Janet's teacher, Charcot, had focused on the neurologial bases of hysteria, Janet was concerned to develop a scientific approach to psychopathology as a mental disorder. His theory that mental pathology results from conflict between unconscious and conscious parts of the mind, and that unconscious mental contents may emerge as symptoms with symbolic meanings led to a public priority dispute with Sigmund Freud.

EARLY BRITISH PSYCHOLOGY

Although the British had the first scholarly journal dedicated to the topic of psychology--Mind, founded in 1876 by Alexander Bain and edited by George Croom Robertson--it was quite a long while before experimental psychology developed there to challenge the strong tradition of "mental philosophy." The experimental reports that appeared in Mind in the first two decades of its existence were almost entirely authored by Americans, especially G. Stanley Hall and his students (notably Henry Herbert Donaldson) and James McKeen Cattell.

Francis Galton's (1822-1911) anthropometric laboratory opened in 1884. There people were tested on a wide variety of physical (e.g., strength of blow) and percepetual (e.g., visual acuity) attributes. In 1886 Galton was visited by James McKeen Cattell who would later adapt Galton's techniques in developing his own mental testing research program in the United States. Galton was not primarily a psychologist, however. The data he accumulated in the anthropometric laboratory primarily went toward supporting his case for eugenics. To help interpret the mounds of data he accumulated, Galton developed a number of important statistical techniques, including the precursors to the scatterplot and the product-moment correlation coefficient (later perfected by Karl Pearson, 1857-1936).

Soon after, Charles Spearman (1863-1945) developed the correlation-based statistical procedure of factor analysis in the process of building a case for his two-factor theory of intelligence, published in 1901. Spearman believed that people have an inborn level of general intelligence or g which can be crystallized into a specific skill in any of a number of narrow content area (s, or specific intelligence). Laboratory psychology of the kind practiced in Germany and the United States was slow in coming to Britain. Although the philosopher James Ward (1843-1925) urged Cambridge University to establish a psychophysics laboratory from the mid-1870s forward, it was not until the 1891 that they put so much as £50 toward some basic apparatus (Bartlett, 1937). A laboratory was established through the assistance of the physiology department in 1897 and a lectureship in psychology was established which first went to W. H. R. Rivers (1864-1922). Soon Rivers was joined by

C.S. Myers (1873-1946) and William McDougall (1871-1938). This group showed as much interest in anthropology as psychology, going with Alfred Cort Haddon (1855-1940) on the famed Torres Straits expedition of 1898.

In 1901 the Psychological Society was established (which renamed itself the British Psychological Society in 1906), and in 1904 Ward and Rivers co-founded the British Journal of Psychology.

SECOND GENERATION GERMAN PSYCHOLOGY

Würzburg School: In 1896, one of Wundt's former Leipzig laboratory assistants, Oswald Külpe (1862-1915), founded a new laboratory in Würzburg. Külpe soon surrounded himself with a number of younger psychologists, most notably Narziß Ach (1871-1946), Karl Bühler (1879-1963), Ernst Dürr (1878-1913), Karl Marbe (1869-1953), and Henry Jackson Watt (1879-1925). Collectively, they developed a new approach to psychological experimentation that flew in the face of many of Wundt's restrictions. Wundt had drawn a distinction between the old philosophical style of self-observation (Selbstbeobachtung) in which one introspected for extended durations on higher thought processes and inner-perception (innere Wahrnehmung) in which one could be immediately aware of a momentary sensation, feeling, or image (Vorstellung). The former was declared to be impossible by Wundt, who argued that higher thought could not be studied experimentally through extended instrospection, but only humanistically through Völkerpsychologie (folk psychology). Only the latter was a proper subject for experimentation.

The Würzburgers, by contrast, designed experiments in which the experimental subject was presented with a compliex stimulus (e.g., a Nietzschean aphorism or a logical problem) and after processing it for a time (e.g., interpreting the aphorism or solving the problem), retrospectively reported to the experimenter all that had passed through his consciousness during the interval. In the process, the Würzburgers claimed to have discovered a number of new elements of consciousness (over and above Wundt's sensations, feelings, and images) including Bewußtseinslagen (conscious sets), Bewußtheiten (awarenesses), and Gedanken (thoughts). In the English-language literature, these are often collectively termed "imageless thoughts," and the debate between Wundt and the Würzburgers as the "imageless thought controversy."

Wundt referred to the Würzburgers' studies as "sham" experiments and criticized them vigorously. Wundt's most significant English student, Edward Bradford Titchener, then working at Cornell, intervened in the dispute, claiming to have conducted extended introspective studies in which he was able to resolve the Würzburgers imageless thoughts into sensations, feelings, and images. He thus, paradoxically, used a method of which Wundt did not approve in order to affirm Wundt's view of the situation.

The imageless thought debate is often said to have been instrumental in undermining the legitimacy of all introspective methods in experimental psychology and, ultimately, in bringing about the behaviorist revolution in American psychology. It was not without its own delayed legacy, however. Herbert Simon cites the work of one Würzburg psychologists in particular, Otto Selz (1881-1943), for having inspired him to develop his famous problem-solving computer algorithms (e.g., Logic Theorist and General Problem Solver) and his "thinking out loud" method for protocol analysis. In addition, Karl Popper studied psychology under Bühler and Selz, and appears to have brought some of their influence, unattributed, to his philosophy of science (Ter Hark, 2004).

Gestalt Psychology: Whereas the Würzburgers debated with Wundt mainly on matters of method, another German movement, centered in Berlin, took issue with the widespread assumption that the aim of psychology should be to break consciousness down into putative basic elements. Instead, they argued that the psychological "whole" has priority and that the "parts" are defined by the structure of the whole, rather than vice versa. Thus, the school was named Gestalt, a German term meaning approximately "form" or "configuration." It was led by Max Wertheimer (1880-1943), Wolfgang Köhler (1887-1967), and Kurt Koffka (1886-1941). Wertheimer had been a student of Austrian philosopher, Christian von Ehrenfels (1859-1932), who claimed that in addition to the sensory elements of a perceived object, there is an extra element which, though in some sense derived from the organization of the standard sensory elements, is also to be regarded as being an element in its own right. He called this extra element Gestalt-qualität or "form-quality." For instance, when one hears a melody, one hears the notes plus something in addition to them which binds them together into a tune--the Gestalt-qualität. It is the presence of this Gestalt-qualität which, according to Von Ehrenfels, allows a tune to be transposed to a new key, using completely different notes, but still retain its identity. Wertheimer took the more radical line that "what is given me by the melody does not arise... as a secondary process from the sum of the pieces as such. Instead, what takes place in each single part already depends upon what the whole is," (1925/1938). In other words, one hears the melody first and only then may perceptually divide it up into notes. Similarly in vision, one sees the form of the circle first--it is given "im-mediately" (i.e. its apprehension is not mediated by a process of part-summation). Only after this primary apprehension might one notice that it is made up of lines or dots or stars.

Gestalt-Theorie was officially initiated in 1912 in an article by Wertheimer on the phi-phenomenon; a perceptual illusion in which two stationary but alternately flashing lights appear to be a single light moving from one location to another. Contrary to popular opinion, his primary target was not behaviorism, as it was not yet a force in psychology. The aim of his criticism was, rather, the atomistic psychologies of Hermann von Helmholtz (1821-1894), Wilhelm Wundt (1832-1920), and other European psychologists of the time.

The two men who served as Wertheimer's subjects in the phi experiment were Köhler and Koffka. Köhler was an expert in physical acoustics, having studied under physicist Max Planck (1858-1947), but had taken his degree in psychology under Carl Stumpf (1848-1936). Koffka was also a student of Stumpf's, having studied movement phenomena and psychological aspects of rhythm. In 1917 Köhler (1917/1925) published the results of four years of research on learning in chimpanzees. Köhler showed, contrary to the claims of most other learning theorists, that animals can learn by "sudden insight" into the "structure" of a problem, over and above the associative and incremental manner of learning that Ivan Pavlov (1849-1936) and Edward Lee Thorndike (1874-1949) had demonstrated with dogs and cats, respectively.

The terms "structure" and "organization" were focal for the Gestalt psychologists. Stimuli were said to have a certain structure, to be organized in a certain way, and that it is to this structural organization, rather than to individual sensory elements, that the organism responds. When an animal is conditioned, it does not simply respond to the absolute properties of a stimulus, but to its properties relative to its surroundings. To use a favorite example of Köhler's, if conditioned to respond in a certain way to the lighter of two gray cards, the animal generalizes the relation between the two stimuli rather than the absolute properties of the conditioned stimulus: it will respond to the lighter of two cards in subsequent trials even if the darker card in the test trial is of the same intensity as the lighter one in the original training trials.

In 1921 Koffka published a Gestalt-oriented text on developmental psychology, Growth of the Mind. With the help of American psychologist Robert Ogden, Koffka introduced the Gestalt point of view to an American audience in 1922 by way of a paper in Psychological Bulletin. It contains criticisms of then-current explanations of a number of problems of perception, and the alternatives offered by the Gestalt school. Koffka moved to the United States in 1924, eventually settling at Smith College in 1927. In 1935 Koffka published his Principles of Gestalt Psychology. This textbook laid out the Gestalt vision of the scientific enterprise as a whole. Science, he said, is not the simple accumulation of facts. What makes research scientific is the incorporation of facts into a theoretical structure.

The goal of the Gestaltists was to integrate the facts of inanimate nature, life, and mind into a single scientific structure. This meant that science would have swallow not only what Koffka called the quantitative facts of physical science but the facts of two other "scientific categories": questions of order and questions of Sinn, a German word which has been variously translated as significance, value, and meaning. Without incorporating the meaning of experience and behavior, Koffka believed that science would doom itself to trivialities in its investigation of human beings.

Having survived the onslaught of the Nazis up to the mid-1930s, all the core members of the Gestalt movement were forced out of Germany to the United States by 1935 (Henle, 1984). Köhler published another book, Dynamics in Psychology, in 1940 but thereafter the Gestalt movement suffered a series of setbacks. Koffka died in 1941 and Wertheimer in 1943. Wertheimer's long-awaited book on mathematical problem-solving, Productive Thinking was published posthumously in 1945 but Köhler was now left to guide the movement without his two long-time colleagues.

THE EMERGENCE OF BEHAVIORISM IN AMERICA

As a result of the conjunction of a number of events in the early 20th century, behaviorism gradually emerged as the dominant school in American psychology. First among these was the increasing skepticism with which many viewed the concept of consciousness: although still considered to be the essential element separating psychology from physiology, its subjective nature and the unreliable introspective method it seemed to require, troubled many. William James' 1904 Journal of Philosophy... article "Does Consciousness Exist?", laid out the worries explicitly.

Second was the gradual rise of a rigorous animal psychology. In addition to Edward Lee Thorndike's work with cats in puzzle boxes in 1898, the start of research in which rats learn to navigate mazes was begun by Willard Small (1900, 1901 in American Journal of Psychology). Robert M. Yerkes's 1905 Journal of Philosophy... article "Animal Psychology and the Criteria of the Psychic" raised the general question of when one is entitled to attribute consciousness to an organism. The following few years saw the emergence of John Broadus Watson (1878-1959) as a major player, publishing his dissertation on the relation between neurological development and learning in the white rat (1907, Psychological Review Monograph Supplement; Carr & Watson, 1908, J. Comparative Neurology & Psychology). Another important rat study was published by Henry H. Donaldson (1908, J. Comparative Neurology &

Psychology). The year 1909 saw the first English-language account of Ivan Pavlov's studies of conditioning in dogs (Yerkes & Morgulis, 1909, Psychological Bulletin).

A third factor was the rise of Watson to a position of significant power within the psychological community. In 1908, Watson was offered a junior position at Johns Hopkins by James Mark Baldwin. In addition to heading the Johns Hopkins department, Baldwin was the editor of the influential journals, Psychological Review and Psychological Bulletin. Only months after Watson's arrival, Baldwin was forced to resign his professorship due to scandal. Watson was suddenly made head of the department and editor of Baldwin's journals. He resolved to use these powerful tools to revolutionize psychology in the image of his own research. In 1913 he published in Psychological Review the article that is often called the "manifesto" of the behaviorist movement, "Psychology as the Behaviorist Views It." There he argued that psychology "is a purely objective experimental branch of natural science", "introspection forms no essential part of its methods..." and "The behaviorist... recognizes no dividing line between man and brute". The following year, 1914, his first textbook, Behavior went to press. Although behaviorism took some time to be accepted as a comprehensive approach, (in no small part because of the intervention of World War I), by the 1920s Watson's revolution was well underway. The central tenet of early behaviorism was that psychology should be a science of behavior, not of the mind, and rejected internal mental states such as beliefs, desires, or goals. Watson himself, however, was forced out of Johns Hopkins by scandal in 1920. Although he continued to publish during the 1920s, he eventually moved on to a career in advertising.

Among the behaviorists who continued on, there were a number of disagreements about the best way to proceed. Neo-behaviorists such as Edward C. Tolman, Edwin Guthrie, Clark L. Hull, and B. F. Skinner debated issues such as: (1) whether to reformulate the traditional psychological vocabulary in behavioral terms or discard it in favor of a wholly new scheme, (2) whether learning takes place all at once or gradually, (3) whether biological drives should be included in the new science in order to provide a "motivation" for behavior, and (4) to what degree any theoretical framework is required over and above the measured effects of reinforcement and punishment on learning. By the late 1950s, Skinner's formulation had become dominant, and it remains a part of the modern discipline under the rubric of Behavior Analysis.

Behaviorism was the ascendant experimental model for research in psychology for much of the 20th century, largely due to the creation and successful application (not least of which in advertising) of conditioning theories as scientific models of human behaviour.

Cognitivism: Noam Chomsky's review of Skinner's book Verbal Behavior (that aimed to explain language acquisition in a behaviorist framework) is considered one of the major theoretical challenges to the type of radical behaviorism that Skinner taught. Chomsky showed that language could not be learned solely from the sort of operant conditioning that Skinner postulated. Chomsky's argument was that as people could produce an infinite variety of sentences unique in structure and meaning, and that these could not possibly be generated solely through experience of natural language. As an alternative, he concluded that there must be internal mental structures-states of mind of the sort that behaviorism rejected as illusory. Similarly, work by Albert Bandura showed that children could learn by social observation, without any change in overt behaviour, and so must be accounted for by internal representations.

The rise of computer technology also promoted the metaphor of mental function as information processing. This, combined with a scientific approach to studying the mind, as well as a belief in internal mental states, led to the rise of cognitivism as the dominant model of the mind.

Links between brain and nervous system function were also becoming common, partly due to the experimental work of people like Charles Sherrington and Donald Hebb, and partly due to studies of people with brain injury. With the development of technologies for accurately measuring brain function, neuropsychology and cognitive neuroscience have become some of the most active areas in contemporary psychology.

With the increasing involvement of other disciplines (such as philosophy, computer science, and neuroscience) in the quest to understand the mind, the umbrella discipline of cognitive science has been created as a means of focusing such efforts in a constructive way.

Dissenting Schools: Not all psychologists, however, have been happy with what they perceive as mechanical models of the mind and human nature. Carl Jung, a one-time follower and contemporary of Freud, was instrumental in introducing notions of spirituality into Freudian psychoanalysis (Freud had rejected religion as a mass delusion).

Alfred Adler, after a brief association with Freud's discussion circle, left to form his own discipline, called Individual (indivisible) Psychology. His influence on contemporary psychology has been considerable, with many approaches borrowing fragments of his theory. A recent rebirth of his legacy, Classical Adlerian Psychology, combines Adler's original theory of personality, style of psychotherapy, and philosophy of living, with Abraham Maslow's vision of optimal functioning.

Humanistic psychology emerged in the 1950s and has continued as a reaction to positivist and scientific approaches to the mind. It stresses a phenomenological view of human experience and seeks to understand human beings and their behavior by conducting qualitative research. The humanistic approach has its roots in existentialist and phenomenological philosophy and many humanist psychologists completely reject a scientific approach, arguing that trying to turn human experience into measurements strips it of all meaning and relevance to lived existence. Some of the founding theorists behind this school of thought are Abraham Maslow, who formulated a hierarchy of human needs; Carl Rogers, who created and developed client centred therapy; and Fritz Perls, who helped create and develop Gestalt therapy.

A further development of Humanistic psychology emerging in the 1970s was Transpersonal psychology, which studies the spiritual dimension of humanity, looking at the possibilities for development beyond the normal ego-boundaries.

3

Researches in Psychology

QUANTITATIVE PSYCHOLOGICAL RESEARCH

Quantitative psychological research is psychological research which performs statistical estimation or statistical inference. This definition distinguishes it from qualitative psychological research; however, many psychologists do not acknowledge any real difference between quantitative and qualitative research. The validity of the distinction is discussed in the article about qualitative psychological research.

Quantitative methods are used in many social sciences.

In psychology qualitative research has come to be defined as research whose findings are not arrived at by statistical or other quantitative procedures. This definition, however, is entirely negative, describing qualitative research by what it lacks rather than by what it possesses. Absence of a characteristic is not itself a defining characteristic-journalism, for example, does not estimate parameters or test null hypotheses, but one would scarcely claim that it is a form of psychological research. If qualitative research has some goal other than estimating parameters or testing, the important issue is what that goal is. Qualitative psychological researchers have described other characteristics of qualitative research which they believe also distinguish it from so-called quantitative psychological research.

Goals Attributed to Qualitative Psychological Research: To many researchers the goal of qualitative psychological research is to develop theory. The crucial question here is the definitions of hypothesis or theory. If what is meant are hypotheses or theories which can be tested by quantitative statistical methods, then the definition is operational and objective. However, it is not universally accepted. Many practitioners reject quantitative methods outright. Other goals have therefore been proposed for qualitative research.

Qualitative research is often said to be naturalistic. That is, its goal is to understand behaviour in a natural setting.

Two other goals attributed to qualitative research are understanding a phenomenon from the perspective of the research participant and understanding the meanings people give to their experience. It attempts to do this by using so-called naturalistic methods-interviewing, observation, ethnography, participant observation and focus groups. Each of these methods seeks to understand the perspective of the research participant within the context of their everyday life. This means that the researcher is concerned with asking broad questions that allow the respondent to answer in their own words. These methods allow the researcher to try to qualify their understanding during the research process through further probing questions. In addition, a method such as observation allows the researcher to observe people within natural settings-particularly those in public places. This has resulted in greater understanding of people's behaviours in for example-lifts, public transport, and queues.

Qualitative research is sometimes said to have as its goal the understanding of the sample studied, rather than generalizing from the sample to the population. However, the results of qualitative research can be applied to other settings-as long as the reader of the research understands the limitations. For example, the research findings of a qualitative case study of primary school children in a particular school and their mobile phone use will tell us more about the mobile phone of children in the general population, than of adults. However, the type of school (public or private), where it was located, and the socio-economic background of the students need to be taken into consideration when applying any findings to other settings (either schools or the general population of children).

In addition to the methods for collecting data mentioned above, qualitative research includes a wide range of ways to analyse the data. One of the most popular of these is known as grounded theory. Others include conversation analysis, discourse analysis, and thematic analysis. In quantitative research these methods are used-as data collection tools rather than as analystic ones.

Qualitative psychological research emphasizes fieldwork, and this emphasis has been offered as a distinguishing mark. However, quantitative researchers often perform fieldwork.

Qualitative psychological research is also described as holistic. That is, qualitative researchers believe in studying phenomena in its context rather than concentrating on narrow aspects of the phenomena. This means that they either observe or participate in the phenomena they are studying, e. g. attending a football game to understand the behaviours of fan, and/or they ask open-ended questions about the behaviour of fans at football games. These questions are holistic because they are designed to understand the context of behaviour-they will usually follow a pattern that replicates the experience eg. what did you do when you arrived? who did you come with? what did you do then? However, similar methods are used by quantitative researchers.

Methods: The philosophical bases of qualitative psychological research are found in phenomenology, ethnomethodology, and naturalistic behaviourism. Its research methods are derived from ethnography and anthropology.

In psychology, the research methods commonly classified as qualitative include:

- participant observation
- direct observation
- unstructured interviewing
- case studies
- content analysis
- protocol analysis
- focus groups

The data collected by researchers using these techniques consist of:

- the results of open-ended interviews
- notes of direct observation
- written documents (answers to questionnaires, diaries, program records, and so on)

After collecting data qualitative psychological researchers goal are to examine their data in depth and in detail.

Most psychological researchers probably use both types of method. In particular, qualitative methods are widely used as exploratory methods; the results of qualitative analysis are used to design quantitative research which tests null hypotheses derived from the qualitative observations. Many psychological researchers prefer qualitative research. They argue that statistically-based research has limitations because it is less able to take into consideration the context of behaviour.

Qualitative researchers have developed their own criteria for assessing reliability and validity. The work of Yvonna Lincoln and Egon Guba is an example of this.

Confirmability is a qualitative concept analogous to the concept of objectivity in quantitative research. It is the degree to which research results can be confirmed by other researchers.

Transferability has been proposed as a qualitative substitute for psychometric validity. Research findings are transferable to the extent to which they can be generalized to settings other than the one in which they were made.

It could be argued, however, that any concept which attempts to assess degree or extent is inherently quantitative.

Status in Psychology: The prevailing opinion in psychology is probably that both approaches offer important benefits, that rejecting one or the other means renouncing some of those benefits, and that the most useful debate is about the circumstances in which the two approaches may most profitably be used.

BIOLOGICAL PSYCHOLOGY

Biological psychology, sometimes referred to as psychobiology or biopsychology, is a subfield of psychology.

Psychology may be defined as the scientific study of mental processes and behavior. Biological psychologists most commonly use an experimental approach to the study of psychology; by definition, the experimental manipulation is usually a biological one. Alternatively, biological psychologists may be interested in measuring some biological variable (e.g., an anatomical, physiological, or genetic variable) in an attempt to relate it quantitatively or qualitatively to a psychological (often behavioral) variable.

Relationship to Psychology and Biology

In many cases, humans may serve as experimental subjects in biological psychology experiments; however, a great deal of the experimental literature in biological psychology comes from the study of non-human species, most frequently rats, mice, and monkeys. As a result, a critical assumption in biological psychology is that organisms share biological and behavioral similarities, enough to permit extrapolations across species. This allies biological psychology closely with comparative psychology, evolutionary psychology, and evolutionary biology. Biological psychology also has paradigmatic and methodological similarities to neuropsychology, which relies heavily on the study of the behavior of humans with nervous system dysfunction (i.e., a non-experimentally based biological manipulation).

Synonyms for biological psychology include biopsychology, behavioral neuroscience, and psychobiology. Physiological psychology is another term often used synonymously with biological psychology, though some authors would make physiological psychology a subfield of biological psychology, with an appropriately more narrow definition.

Research Methods of Biological Psychology

The distinguishing characteristic of a biological psychology experiment is that either the independent variable of the experiment is biological, or some dependent variable is biological. In other words, the nervous system of the organism under study is permanently or temporarily altered, or some aspect of the nervous system is measured (usually to be related to a behavioral variable). Most commonly, these manipulations and measurements concern non-human subjects.

Disabling or Decreasing Neural Function

- *Lesions:* A classic method in which a brain-region of interest is disabled. Lesions can be placed with relatively high accuracy thanks to a variety of brain 'atlases' which provide a map of brain regions in 3-dimensional stereotactic coordinates.
  - Electrolytic lesions-Neural tissue is destroyed by the use of electric current.
  - Chemical lesions-Neural tissue is destroyed by the infusion of a neurotoxin.
  - Temporary lesions-Neural tissue is temporarily disabled by cooling or by the use of anesthetics such as tetrodotoxin.
  - Transcranial magnetic stimulation-A new technique usually used with human subjects in which a magnetic coil applied to the scalp causes unsystematic electrical activity in nearby cortical neurons which can be experimentally analyzed as a functional lesion.
- *Psychopharmacological Manipulations:* A chemical receptor antagonist reduces neural activity by interfering with neurotransmission. Antagonists can be delivered systemically (such as by intravenous injection) or locally (intracebrally) during a surgical procedure.

Enhancing Neural Function

- *Electrical Stimulation:* A classic method in which neural activity is enhanced by application of a small electrical current (too small to cause significant cell death).
- *Psychopharmacological Manipulations:* A chemical receptor agonist facilitates neural activity by enhancing or replacing endogenous neurotransmitters. Agonists can be delivered systemically (such as by intravenous injection) or locally (intracebrally) during a surgical procedure.

- *Transcranial Magnetic Stimulation:* In some cases (for example, studies of motor cortex, this technique can be analyzed as having a stimulatory effect (rather than as a functional lesion).

Measuring Neural Activity

- *Single Unit Recording:* The measurement of the electrical activity of one neuron, often in the context of an ongoing behavioral (psychological) task.
- *Multielectrode Recording:* The use of a bundle of fine electrodes to record the simultaneous activity of up to hundreds of neurons.
- *fMRI:* Functional magnetic resonance imaging, a technique most frequently applied on human subjects, in which changes in cerebral blood flow can be detected in an MRI apparatus and are taken to indicate relative activity of larger scale brain regions (i.e., on the order of hundreds of thousands of neurons).
- *Electroencephalography:* Or EEG; and the derivative technique of event-related potentials, in which scalp electrodes monitor the average activity of neurons in the cortex (again, used most frequently with human subjects).
- *Functional Neuroanatomy:* In which the expression of some anatomical marker is taken to reflect neural activity. For example, the expression of immediate early genes is thought to be caused by vigorous neural activity. Likewise, the injection of 2-deoxyglucose prior to some behavioral task can be followed by anatomical localization of that chemical; it is taken up by neurons that are electrically active.

Genetic Manipulations

- *QTL Mapping:* The influence of a gene in some behavior can be statistically inferred by studying inbred strains of some species, most commonly mice. The recent sequencing of the genome of many species, most notably mice, has facilitated this technique.

- *Selective Breeding:* Organisms, often mice, may be bred selectively among inbred strains to create a recombinant congenic strain. This might be done to isolate an experimentally interesting stretch of DNA derived from one strain on the background genome of another strain to allow stronger inferences about the role of that stretch of DNA.
- *Genetic Engineering:* The genome may also be experimentally-manipulated; for example, knockout mice can be engineered to lack a particular gene, or a gene may be expressed in an strain which does not normally do so (the 'knock in'). Advanced techniques may also permit the expression or suppression of a gene to occur by injection of some regulating chemical.

Topic Areas in Biological Psychology

In general, biological psychologists study the same issues as academic psychologists, though limited by the need to use nonhuman species. As a result, the bulk of literature in biological psychology deals with mental processes and behaviors that are shared across mammalian species, such as:

- Sensation and perception
- Motivated behavior (hunger, thirst, sex)
- Control of movement
- Learning and memory
- Sleep and biological rhythms
- Emotion.

However, with increasing technical sophistication and with the development of more precise noninvasive methods that can be applied to human subjects, biological psychologists are beginning to contribute to other classical topic areas of psychology, such as:

- Language
- Reasoning and decision making
- Consciousness.

Biological psychology has also had a strong history of contributing to medical disorders including those that fall under the purview of clinical psychology and psychopathology (also known as abnormal psychology). Although animal models for all mental illnesses do not exist, the field has contributed important therapeutic data on a variety of conditions, including:

- Parkinson's Disease
- Huntington's Disease
- Alzheimer's Disease
- Depression
- Schizophrenia
- Autism
- Anxiety
- Drug abuse, including alcoholism.

Philosophical Implications: Biopsychology is related to the philosophical problem of mental causation through the common premise that all mental and psychological phenomena supervene on material structures and processes. As an alternative to this premise, it has been suggested (for example) that the brain may be sensitive to quantum fluctuations, and that this may exempt it from strict physical and biological determinacy. However, such ideas remain speculative, as is the more widespread assumption that some combination of physical and biological causation accounts for all mental and behavioral phenomena.

COGNITIVE PSYCHOLOGY

Cognitive Psychology is the school of psychology that examines internal mental processes such as problem solving, memory, and language. It had its foundations in the Gestalt psychology of Max Wertheimer, Wolfgang Köhler, and Kurt Koffka, and in the work of Jean Piaget, who studied intellectual development in children. Cognitive psychologists are interested in how people understand, diagnose, and solve problems, concerning themselves with the mental processes which mediate between stimulus and response. Cognitive theory contends that solutions to problems take the form of algorithms-rules that are not necessarily understood but promise a solution, or heuristics-rules that are understood but that do not always guarantee solutions. In other instances, solutions may be found through insight, a sudden awareness of relationships.

Introduction

Ulric Neisser coined the term 'cognitive psychology' in his book published in 1967, wherein Neisser provides a definition of cognitive psychology, emphasising that it is a point of view which postulates the mind as having a certain conceptual structure. Neisser's point of view endows the discipline a scope which expands beyond high-level concepts such as "reasoning", often espoused in other works as a definition of cognitive psychology. Neisser's definition of cognition illustrates this well:

...the term "cognition" refers to all processes by which the sensory input is transformed, reduced, elaborated, stored, recovered, and used. It is concerned with these processes even when they operate in the absence of relevant stimulation, as in images and hallucinations... Given such a sweeping definition, it is apparent that cognition is involved in everything a human being might possibly do; that every psychological phenomenon is a cognitive phenomenon. But although cognitive psychology is concerned with all human activity rather than some fraction of it, the concern is from a particular point of view. Other viewpoints are equally legitimate and necessary. Dynamic psychology, which begins with motives rather than with sensory input, is a case in point. Instead of asking how a man's actions and experiences result from what he saw, remembered, or believed, the dynamic psychologist asks how they follow from the subject's goals, needs, or instincts.

Cognitive psychology is radically different from previous psychological approaches in two key ways.

- It accepts the use of the scientific method, and generally rejects introspection as a valid method of investigation, unlike phenomenological methods such as Freudian psychology.
- It explicitly acknowledges the existence of internal mental states (such as belief, desire and motivation) unlike behaviorist psychology.

The school of thought arising from this approach is known as cognitivism.

Cognitive psychology is one of the more recent additions to psychological research, having only developed as a separate area within the discipline since the late 1950s and early 1960s (though there are examples of cognitive thinking from earlier researchers). The cognitive approach was brought to prominence by Donald Broadbent's book Perception and Communication in 1958. Since that time, the dominant paradigm in the area has been the information processing model of cognition that Broadbent put forward. This is a way of thinking and reasoning about mental processes, envisaging them as software running on the computer that is the brain. Theories commonly refer to forms of input, representation, computation or processing, and outputs. Applied to language as the primary mental knowledge representation system, cognitive psychology has exploited tree and network mental models. Its singular contribution to AI and psychology in general is the notion of a semantic network. One of the first cognitive psychologists, George Miller is well-known for dedicating his career to the development of WordNet, a semantic network for the English language. Development began in 1985 and is now the foundation for many machine ontologies.

This way of conceiving mental processes has pervaded psychology more generally over the past few decades, and it is not uncommon to find cognitive theories within social psychology, personality, abnormal psychology, and developmental psychology; the application of cognitive theories to comparative psychology has driven many recent studies in animal cognition. The information processing approach to cognitive functioning is currently being questioned by new approaches in psychology, such as dynamical systems, and the embodiment perspective.

Because of the use of computational metaphors and terminology, cognitive psychology was able to benefit greatly from the flourishing of research in artificial intelligence and other related areas in the 1960s and 1970s. In fact, it developed as one of the significant aspects of the inter-disciplinary subject of cognitive science, which attempts to integrate a range of approaches in research on the mind and mental processes.

COMPARATIVE PSYCHOLOGY

Comparative psychology, taken in its most usual, broad sense, refers to the study of the behavior and mental life of animals other than human beings. It is synonymous with animal psychology, but although the latter would be a more accurate term, it is less often used. Veterinarians sometimes use the phrase "animal psychology" to refer specifically to the study of disordered behaviour in animals, discussed below.

History: Comparative psychology may be said to have come into being in the late nineteenth century, with the work of George Romanes on animal psychology, inspired by Charles Darwin. It was confirmed as an important discipline within academic psychology by the experiments on instrumental learning of Edward L. Thorndike, and on classical conditioning by Ivan Pavlov. Early comparative psychology experimented on animals to discover fundamental principles, especially of learning, that might be applicable to humans. Interest in the social behavior of animals has always also been a lesser part of comparative psychology; however the famous studies of social dominance in chickens by T. Schjelderup-Ebbe, giving rise to the concept of the "pecking order", were an important early contribution to the field.

Comparative Psychology and the Comparative Method: Strictly speaking, comparative psychology ought to involve the use of a comparative method, in which similar studies are carried out on animals of different species, and the results interpreted in terms of their different phylogenetic or ecological backgrounds. Throughout the long history of comparative psychology, repeated attempts have been made to enforce this more disciplined approach, especially since the rise of ethology in the mid twentieth century. Behavioral ecology in the 1970s gave a more solid base of knowledge against which a true comparative psychology could develop. However, the broader use of the term "comparative psychology" is enshrined in the names of learned societies and academic journals, not to mention in the minds of psychologists of other specialisms, so it is never likely to disappear completely.

A persistent question with which comparative psychologists have been faced is the relative intelligence of different species of animal. Much effort has gone into explaining that this may not be a good question, but it will not go away. Indeed, some early attempts at a genuinely comparative psychology involved evaluating how well animals of different species could learn different tasks. These attempts foundered; in retrospect it can be seen that they were not sufficiently sophisticated, either their analysis of the demands of different tasks, or in their choice of species to compare. More recent comparative work has been more successful, partly because it has drawn upon studies in ethology and behavioral ecology to make informed choices of species and tasks to compare.

Species Studied: A wide variety of species have been studied by comparative psychologists. However a small number have dominated the scene. Pavlov's early work used dogs, but although they have been the subject of occasional studies since then they have not figured prominently. Increasing interest in the study of abnormal animal behaviour has led to a return to the study of most kinds of domestic animal. Thorndike began his studies with cats, but American comparative psychologists quickly shifted to the more economical rat, which remained the almost invariable subject for the first half of the twentieth century and continues to be used. Skinner introduced the use of pigeons, and they continue to be important in some fields. There has always been interest in studying various species of primate; important contributions to social and developmental psychology were made by Harry F. Harlow's studies of maternal deprivation in rhesus monkeys. Interest in primate studies has increased with the rise in studies of animal cognition. Other animals thought to be intelligent have also been increasingly studied. Examples include various species of corvid, parrots-especially the African Gray Parrot-and dolphins.

Animal Cognition: Since the 1990s, comparative psychology has undergone a reversal in its fundamental approach. Instead of seeking principles in animal behaviour in order to explain human performance, comparative psychologists started taking principles that have been uncovered in the study of human cognition and testing them in animals of other species. This approach is referred to as the study of animal cognition. It has led to significant advances in our understanding of concept formation, memory, problem solving and other cognitive abilities in animals.

Disorders of Animal Behaviour: Today an animal's psychological constitution is recognised by veterinary surgeons as an important part of its living conditions in domestication or captivity.

Common causes of disordered behaviour in captive or pet animals are lack of stimulation, inappropriate stimulation, or overstimulation. These conditions can lead to disorders, unpredictable and unwanted behaviour, and sometimes even physical symptoms and diseases. For example, rats that are exposed to loud music for a long period will ultimately develop unwanted behaviours that have been compared with human psychosis, like biting their owners.

The way dogs behave when understimulated is widely believed to depend on the breed as well as on the individual animal's character. For example, huskies have been known to completely ruin gardens and houses, if they are not allowed enough activity. Dogs are also prone to psychological damage if they are subjected to violence. If they are treated very badly, they may become dangerous.

The systematic study of disordered animal behaviour draws on research in comparative psychology, including the early work on conditioning and instrumental learning, but also on ethological studies of natural behaviour. However, at least in the case of familiar domestic animals, it also draws on the accumulated experience of those who have worked closely with the animals.

PSYCHOLOGICAL CHANGES IN HUMAN DEVELOPMENT

Developmental psychology, also known as Human Development, is the scientific study of progressive psychological changes that occur in human beings as they age. Originally concerned with infants and children, and later other periods of great change such as adolescence and aging, it now encompasses the entire life span. This field examines change across a broad range of topics including motor skills and other psycho-physiological processes, problem solving abilities, conceptual understanding, acquisition of language, moral understanding, and identity formation. Developmental psychologists investigate key questions, such as whether children are qualitatively different from adults or simply lack the experience that adults draw upon. Other issues that they deal with is the question of whether development occurs through the gradual accumulation of knowledge or through shifts from one stage of thinking to another; or if children are born with innate knowledge or figure things out through experience; and whether development is driven by the social context or by something inside each child.

Developmental psychology informs several applied fields, including: educational psychology, child psychopathology and developmental forensics. Developmental psychology complements several other basic research fields in psychology including social psychology, cognitive psychology, cognitive development, and comparative psychology.

Psychological Theory of Development

Many theoretical perspectives attempt to explain development, among the most prominent are: Jean Piaget's Stage Theory, Lev Vygotsky's Social Contextualism (and its heir, the Ecological Systems Theory of Urie Bronfenbrenner), and especially the information processing framework employed by cognitive psychology.

Historical theories continue to provide a basis for additional research, among them are Erik Erikson's eight stages of psychosocial development and John B. Watson's and B. F. Skinner's Behaviorism. Many other theories are prominent for their contributions to particular aspects of development. For example, Attachment theory describes kinds of interpersonal relationships and Lawrence Kohlberg describes stages in moral reasoning.

Ecological Systems Theory: Generally regarded as one of the world's leading scholars in the field of developmental psychology, Bronfenbrenner's primary contribution was his Ecological Systems Theory, in which he delineated four types of nested systems, with bi-directional influences within and between systems.

1. *Microsystem:* Immediate environments (family, school, peer group, neighborhood, and childcare environments)
2. *Mesosystem:* A system comprised of connections between immediate environments (i.e., a child's home and school)
3. *Exosystem:* External environmental settings which only indirectly affect development (such as parent's workplace)
4. *Macrosystem:* The larger cultural context (Eastern vs. Western culture, national economy, political culture, subculture)

Each system contains factors that can powerfully shape development, as can the interaction of factors across systems.

The major statement of this theory, The Ecology of Human Development (1979) had widespread influence on the way psychologists and others approached the study of human beings and their environments. It has been said that before Bronfenbrenner, child psychologists studied the child, sociologists examined the family, anthropologists the society, economists the economic framework of the times and political scientists the structure. As a result of Bronfenbrenner's groundbreaking work in "human ecology," a field that he created, these environments-from the family to economic and political structures-were viewed as part of the life course from childhood to adulthood.

Role of Experience: A significant question in developmental psychology is the relation between innateness and environmental influence in regard to any particular aspect of development. This is often referred to as "nature versus nurture" or nativism versus empiricism. A nativist account of development would argue that the processes in question are innate, that is, they are specified by the organism's genes. An empiricist perspective would argue that those processes are acquired in interaction with the environment. Today developmental psychologists rarely take such extreme positions with regard to most aspects of development; rather they investigate, among many other things, the relationship between innate and environmental influences. One of the ways in which this relationship has been explored in recent years is through the emerging field of evolutionary developmental psychology. One area where this innateness debate has been prominently portrayed is in research on language acquisition. A major question in this area is whether or not certain properties of human language are specified genetically or can be acquired through learning. The nativist position argues that the input from language is too impoverished for infants and children to acquire the structure of language. Linguist Noam Chomsky asserts that, evidenced by the lack of sufficient information in the language input, there is a universal grammar that applies to all human languages and is pre-specified. This has led to the idea that there is a special cognitive module suited for learning language, often called the language acquisition device.

The empiricist position on the issue of language acquisition suggests that the language input does provide the necessary information required for learning the structure of language and that infants acquire language through a process of statistical learning. From this perspective, language can be acquired via general learning methods that also apply to other aspects of development, such as perceptual learning. There is a great deal of evidence for components of both the nativist and empiricist position, and this is a hotly debated research topic in developmental psychology.

On the other hand, Chomsky's critique of a specific empiricist position on this issue, radical behaviorist Burrhus Frederic Skinner's Verbal Behavior written in 1957, is widely considered among developmental psychologists to have sparked the decline in influence of behaviorism and signaled the beginning of the cognitive revolution in psychology.

Mechanisms of Development: Developmental psychology is concerned not only with describing the characteristics of psychological change over time, but also seeks to explain the principles and internal workings underlying these changes. Understanding these factors is aided by the use of models. Developmental models are often computational, but they do not necessarily need to be. A model must simply account for the means by which a process takes place. This is sometimes done in reference to changes in the brain that may correspond to changes in behavior over the course of the development.

Computational accounts of development often use either symbolic, connectionist (neural network), or dynamical systems models to explain the mechanisms of development.

History of Developmental Psychology

The modern form of developmental psychology has its roots in the rich psychological tradition represented by Heraclitus, Aristotle and Descartes. William Shakespeare had his melancholy character Jacques (in As You Like It) articulate the seven ages of man: these included three stages of childhood and four of adulthood. In the mid-eighteenth century Jean Jacques Rousseau described three stages of childhood: infans (infancy), puer (childhood) and adolescence in Emile: Or, On Education. Rousseau's ideas were taken up strongly by educators at the time.

In the late nineteenth century, psychologists familiar with the evolutionary theory of Darwin began seeking an evolutionary description of psychological development; prominent here was G. Stanley Hall, who attempted to correlate ages of childhood with previous ages of mankind.

A more scientific approach was initiated by James Mark Baldwin, who wrote essays on topics that included Imitation: A Chapter in the Natural History of Consciousness and Mental Development in the Child and the Race: Methods and Processes. In 1905, Sigmund Freud articulated five psychosexual stages. Later, Rudolf Steiner articulated stages of psychological development throughout human life. The first three of these stages, which correspond closely with Piaget's later-described stages of childhood, were first presented in Steiner's 1911 essay The Education of the Child; his descriptions have been taken up by educators (in the Waldorf Schools) and by psychologists. By the early to mid-twentieth century, the work of Vygotsky and Piaget, mentioned above, had established a strong empirical tradition in the field.

The Role of Mothers: Traditionally mothers (and women generally) were emphasized to the exclusion of other caregivers. This has begun to change, with the emphasis now placed on a primary caregiver (regardless of gender or biological relation), as well as all persons directly or indirectly influencing the child (the family system). However, studies are showing that the role of the mother/father are more significant than first thought as we moved into the concept of primary caregiver. For example, the traditional father had little to do with an infant directly, but his method of interacting with the mother (supportive, abusive, neglectful) had a great deal of impact on the infant indirectly.

Stages of Development

Prenatal: The prenatal development of human beings is viewed in three separate stages:

1. Germinal (conception through week 2)
2. Embryonic (weeks 3 through 8)
3. Fetal (week 9 through birth)

These stages are not the same as the trimesters of a woman's pregnancy.

The germinal stage least resembles a grown human. It begins when a sperm penetrates an egg in the act of conception (normally the result of sexual intercourse between a man and a woman). At this point a zygote is formed. Through the process of mitosis the cells divide and double.

The embryonic stage occurs once the zygote has firmly implanted itself into the uterine wall. It is in this stage that the vital organs are formed, and while the external body is still extremely dissimilar from an adult human, some features such as eyes and arms, and eventually ears and feet become recognizable. The fetal period is when the brain most substantially forms, becoming more and more complex over the last few months.

During pregnancy the risk to the developing child from drugs and other teratogens, spousal abuse and other stress on the mother, nutrition and the age of the mother are quite acute. Three methods of determining fetal defects and health include the ultrasound, amniocentesis, and chorionic villus sampling.

Ultrasound uses sound waves and a computer monitor, and is non-invasive, thus minimizing potential harm to fetus and mother. Unfortunately its ability to determine potential defect is also far less comprehensive than more risky methods. Chorionic villus sampling is a form of prenatal diagnosis to determine genetic abnormalities in the fetus. It entails getting a sample of the chorionic villus (placental tissue) and testing it. It is generally carried out only on pregnant women over the age of 35 and those who have a higher risk of Down syndrome and other chromosomal conditions.

The advantage of CVS is that it can be carried out at 10-12 weeks of pregnancy, earlier than amniocentesis (which is carried out at 15-18 weeks). However, it is more risky than amniocentesis, with a 1 in 100 to 200 risk that it will cause a miscarriage.

Amniocentesis is another medical procedure used for prenatal diagnosis, in which a small amount of amniotic fluid is extracted from the amnion around a developing fetus. It is usually offered when there may be an increased risk for genetic conditions (i.e. Down syndrome, sickle-cell disease, cystic fibrosis, etc.) in the pregnancy. Amniocentesis done in the second trimester is often said to have a risk of fetal death between about 1 in 400 and 1 in 200. Often, genetic counseling is done before amniocentesis, or other types of genetic testing, is offered.

Although difficult, some methods of treating fetal disorders have been developed, both surgical and drug based. Genetic testing prior to pregnancy is also increasingly available.

Infancy: From birth until the child begins to speak, they are referred to as an infant. Developmental psychologists vary widely in their assessment of the infant's psychology, and the influence the outside world has upon it, but certain aspects are relatively clear.

While no agreement has yet been reached regarding the level of stimulation an infant requires, we are well aware that a normal level of stimulation is very important, and that a lack of stimulation and affection can result in retardation and a host of other developmental and social disorders. Some feel that classical music, particularly Mozart is good for an infant's mind. While some tentative research has shown it to be helpful to older children, no conclusive evidence is available involving infants.

The majority of an infant's time is spent in sleep. At first this sleep is evenly spread throughout the day and night, but after a couple of months they generally become diurnal. Infants can be seen to have 6 states, grouped into pairs:

- quiet sleep and active sleep (dreaming, when REM occurs)
- quiet waking, and active waking
- fussing and crying

Infants respond to stimuli differently when in these different states. Habituation is frequently used in testing psychological phenomenon. Both infants and adults look less and less as a result of consistent exposure to a particular stimulus. The amount of time spent looking to a presented alternate stimulus (after habituation to the initial stimulus) is indicative of the strength of the remembered percept of the previous stimulus, or dishabituation. Habituation is used to discover the resolution of perceptual systems, for example, by habituating a subject to one stimulus, and then observing responses to similar ones, one can detect the smallest degree of difference that is detectable by the subject.

Infants have a wide variety of reflexes, some of which are permanent (blinking, gagging), and others transient in nature. Some with obvious purposes, some are clearly vestigial and some do not have obvious purposes. Primitive reflexes reappear in adults under certain conditions. Namely, neurological conditions like dementia, traumatic lesions, etc. A partial list of infantile reflexes includes:

- Moro reflex or startle reflex:
  1. Startle
  2. spreading out the arms (abduction)
  3. unspreading the arms (adduction)
  4. Crying (usually)
- Tonic neck reflex or fencer's reflex
- Rooting reflex, sucking reflex, suckling reflex: can be initiated by scratching the infant's cheek; the reaction is pursing of the lips for sucking.

- Stepping reflex, step-up reflex: can be initiated if you support the infant upright from its armpits below a given surface so the baby lifts its foot and steps up on the surface (like climbing a stair).
- Grasp reflex: can be initiated by scratching the infant's palm.
- Parachute reflex the infant is suspended by the trunk and suddenly lowered as if falling for an instant. The child spontaneously throws out the arms as a protective mechanism. The parachute reflex appears before the onset of walking.
- Plantar reflex or Babinski reflex a finger is stroked firmly down the outer edge of the baby's sole; the toes spread and extend out.

Infants have particularly poor vision, and are legally blind. They are capable of sight, however blurry. This improves over time, based on experience. Infants less than 2 months old are also thought to be color blind. Hearing is well-developed prior to birth, however, and a preference for the mother's heartbeat is well established. Infants are fairly good at detecting the direction from which a sound comes, and by 18 months their hearing ability is approximately equal to that of adults.

Smell and taste are present, with infants having been shown to prefer the smell and taste of a banana, while rejecting the taste of shrimp. There is good evidence for infants preferring the smell of their mother to that of others.

Infants have a fully developed sense of touch at birth, and the myth believed by some doctors even today that infants feel no pain is inaccurate. Doctors are slowly becoming aware of the need for pain prevention for newborns.

Piaget felt that there were several sensorimotor stages within his broader Theory of cognitive development.

- The first sub-stage occurs from birth to six weeks and is associated primarily with the development of reflexes. Three primary reflexes are described by Piaget: sucking of objects in the mouth, following moving or interesting objects with the eyes, and closing of the hand when an object makes contact with the palm (palmar grasp). Over these first six weeks of life, these reflexes begin to become voluntary actions; for example, the palmar reflex becomes intentional grasping. (Gruber and Vaneche, 1977).

- The second sub-stage occurs from six weeks to four months and is associated primarily with the development of habits. Primary circular reactions or repeating of an action involving only ones own body begin. An example of this type of reaction would involve something like an infant repeating the motion of passing their hand before their face. Also at this phase, passive reactions, caused by classical or operant conditioning, can begin (Gruber et al., 1977).
- The third sub-stage occurs from four to nine months and is associated primarily with the development of coordination between vision and prehension. Three new abilities occur at this stage: intentional grasping for a desired object, secondary circular reactions, and differentiations between ends and means. At this stage, infants will intentionally grasp the air in the direction of a desired object, often to the amusement of friends and family. Secondary circular reactions, or the repetition of an action involving an external object occur begin; for example, moving a switch to turn on a light repeatedly. The differentiation between means also occurs. This is perhaps one of the most important stages of a child's growth as it signifies the dawn of logic (Gruber et al., 1977). Towards the late part of this sub-stage infants begin to have a sense of object permanence, passing the A-not-B error test.
- The fourth sub-stage occurs from nine to twelve months and is associated primarily with the development of logic and the coordination between means and ends. This is an extremely important stage of development, holding what Piaget calls the "first proper intelligence." Also, this stage marks the beginning of goal orientation, the deliberate planning of steps to meet an objective (Gruber et al. 1977).

- The fifth sub-stage occurs from twelve to eighteen months and is associated primarily with the discovery of new means to meet goals. Piaget describes the child at this juncture as the "young scientist," conducting pseudo-experiments to discover new methods of meeting challenges (Gruber et al. 1977).
- The sixth sub-stage is associated primarily with the beginnings of insight, or true creativity. This marks the passage into the preoperational stage.

When studying infants, the habituation methodology is an example of a method often used to assess their performance. This method allows researchers to obtain information about what types of stimuli an infant is able to discriminate. In this paradigm, infants are habituated to a particular stimulus and are then tested using different stimuli to evaluate discrimination. The critical measure in habituation is the infants' level of interest. Typically, infants prefer stimuli that are novel relative to those they have encountered previously. Several methods are used to measure infants' preference. These include the high-amplitude sucking procedure, in which infants suck on a pacifier more or less depending on their level of interest, the conditioned foot-kick procedure, in which infants move their legs to indicate preference, and the head-turn preference procedure, in which the infant's level of interest is measured by the amount of time spent looking in a particular direction. A key feature of all these methods is that, in each situation, the infant controls the stimuli being presented. This gives researchers a means of measuring discrimination. If an infant is able to discriminate between the habituated stimulus and a novel stimulus, they will show a preference for the novel stimulus. If, however, the infant cannot discriminate between the two stimuli, they will not show a preference for one over the other.

Object permanence is an important stage of cognitive development for infants. Numerous tests regarding it have been done, usually involving a toy, and a crude barrier which is placed in front of the toy, and then removed, repeatedly. In sensorimotor stages 1 and 2, the infant is completely unable to comprehend object permanence. Jean Piaget conducted experiments with infants which led him to conclude that this awareness was typically achieved at eight to nine months of age. Infants before this age are too young to understand object permanence, which explains why infants at this age do not cry when their mothers are gone. "Out of sight, out of mind." A lack of Object Permanence can lead to A-not-B errors, where children reach for a thing at a place where it should not be.

Toddler: Intelligence is demonstrated through the use of symbols, language use matures, and memory and imagination are developed. Thinking is done in a nonlogical, nonreversible manner. Egocentric thinking predominates.

Socially, toddlers are little people attempting to become independent. They walk, talk, use the toilet, and get food for themselves. Self-control begins to develop. If taking the initiative to explore, experiment, risk mistakes in trying new things, and test their limits is encouraged by the caretaker(s) the child will become autonomous, self-reliant, and confident. If the caretaker is overprotective or disapproving of independent actions, the toddler may begin to doubt their abilities and feel ashamed for the desire for independence. The child's autonomic development will be inhibited, be less prepared to successfully deal with the world in the future.

Early Childhood: When children go to preschool, they broaden their social horizons and become more engaged with those around them. Impulses are channeled into fantasies, which leaves the task of the caretaker to balance eagerness for pursuing adventure, creativity and self expression with the development of responsibility. If caretakers are properly encouraging while being consistently disciplinary, children are more likely to develop positive self-esteem while becoming more responsible, and will follow through on assigned activities. If not allowed to decide which activities to perform, children may begin to feel guilt upon contemplating taking initiative. This negative association with independence will lead them to let others make decisions in place of them.

Childhood: In this stage intelligence is demonstrated through logical and systematic manipulation of symbols related to concrete objects. Operational thinking develops, which means actions are reversible, and egocentric thought diminishes.

Children go through the transition from the world at home to that of school and peers. Children learn to make things, use tools, and acquire the skills to be a worker and a potential provider. Children can now receive feedback from outsiders about their accomplishments. If children can discover pleasure in intellectual stimulation, being productive, seeking success, they will develop a sense of competence. If they are not successful or cannot discover pleasure in the process, they may develop a sense of inferiority and feelings of inadequacy that may haunt them throughout life. This is when children think of them selves as industrious or as inferior.

Adolescence: Adolescence is the period of life between the onset of puberty and the full commitment to an adult social role, such as worker, parent, and/or citizen. It is the period known for the formation of personal and social identity and the discovery of moral purpose. Intelligence is demonstrated through the logical use of symbols related to abstract concepts and formal reasoning. A return to egocentric thought often occurs early in the period. Only 35% develop the capacity to reason formally during adolescence or adulthood.

The adolescent asks "Who am I? Who do I want to be?" Like toddlers, adolescents must explore, test limits, become autonomous, and commit to an identity, or sense of self. Different roles, behaviours and ideologies must be tried out to select an identity. Role confusion, inability to choose vocation, sexual orientation and one's role in life can result from a failure to achieve a sense of identity.

Early Adulthood: The person must learn how to form intimate relationships, both in friendship and love. The development of this skill relies on the resolution of other stages. It may be hard to establish intimacy if you haven't developed trust or a sense of identity. If this skill is not learned the alternative is alienation, isolation, a fear of commitment and the inability to depend on others.

Middle Age: Middle adulthood generally refers to the period between ages 35 to 60. During this period, the middle-aged experience a conflict between generativity and stagnation. They may either feel a sense of contributing to the next generation and their community or a sense of purposelessness.

Physically, the middle-aged experience a decline in muscular strength, reaction time, sensory keenness, and cardiac output. Also, women experience menopause and a sharp drop in the hormone estrogen. Men do not have an equivalent to menopause, but they do experience a decline in sperm count and speed of ejaculation and erection. Most men and women remain capable of sexual satisfaction after middle age.

Old Age: This stage generally refers to those over 60 years. During old age, people experience a conflict between integrity vs. despair. When reflecting on their life, they either feel a sense of accomplishment or failure.

Physically, older people experience a decline in muscular strength, reaction time, stamina, hearing, distance perception, and the sense of smell. They also are more susceptible to severe diseases such as cancer and pneumonia due to a weakened immune system. Mental disintegration may also occur, leading to Dementia or Alzheimer's Disease. However, partially due to a lifetime's accumulation of antibodies, the elderly are less likely to suffer from common diseases such as the cold or flu.

Whether or not intellectual powers increase or decrease with age remains controversial. Longitudinal studies have suggested that intellect declines, while cross-sectional studies suggest that intellect is stable. It is generally believed that crystallized intelligence increases up to old age, while fluid intelligence decreases with age.

Schools of Psychology

Cognitive Development: Cognitive development is primarily concerned with the ways in which infants and children acquire and develop internal mental capabilities such as problem solving, memory, and language.. Major topics in cognitive development are the study of language acquisition and the development of perceptual and motor skills. Piaget was one of the influential early psychologists to study the development of cognitive abilities. His theory suggests that development proceeds through a set of stages from infancy to adulthood and that there is an end point or goal. Other accounts, such as that of Lev Vygotsky, have suggested that development does not progress through stages rather that the development process that begins at birth and continues until death is too complex for such structure and finality. Rather, from this viewpoint, developmental processes proceed more continuously, thus it should be analyzed, instead of a product to be obtained.

In addition, modern cognitive development has largely moved away from Piagetian stage theories, and is influenced by accounts of domain specificity, which argue that development is guided by innate evolutionarily specificed and content-specific information processing mechanisms.

Social Development: Social psychology is the study of the nature and causes of human social behavior, with an emphasis on how people think towards each other and how they relate to each other. As the mind is the axis around which social behavior pivots, social psychologists tend to study the relationship between mind(s) and social behaviors. In early-modern social science theory, John Stuart Mill, Comte, and others, laid the foundation for social psychology by asserting that human social cognition and behavior could and should be studied scientifically like any other natural science.

Attachment Theory: Attachment Theory focuses on close, intimate, emotionally meaningful relationships. Its methods of study involve such approaches as the Strange Situation Protocol developed by Mary Ainsworth and the Adult Attachment Interview developed by Mary Main. Attachment Theory was developed by Sir John Bowlby. The attachment is described as a biological system that evolved to ensure the survival of the infant. Attachment behavior is evoked whenever the person is threatened or stressed and involves actions to move toward the person(s) who create a sense of physical, emotional and psychological safety for the individual.

Research Methods

Developmental psychology employs many of the research methods used in other areas of psychology. However, infants and children cannot always be tested in the same ways as adults, so different methods are often used to study development.

Child Research Methods: When studying older children, especially adolescents, adult measurements of behavior can often be used, but they may need to be simplified to allow children to perform certain tasks.

Lifespan Development: Developmental psychologists have a number of methods to study changes in individuals over time.

In a longitudinal study, a researcher observes many individuals born at or around the same time (a cohort) and carries out new observations as members of the cohort age. This method can be used to draw conclusions about which types of development are universal (or normative) and occur in most members of a cohort. Researchers may also observe ways in which development varies between individuals and hypothesize about the causes of variation observed in their data. Longitudinal studies often require large amounts of time and funding, making them unfeasible in some situations. Also, because members of a cohort all experience historical events unique to their generation, apparently normative developmental trends may in fact be universal only to their cohort.

In a cross-sectional study, a researcher observes differences between individuals of different ages at the same time. This generally requires less resources than the longitudinal method, and because the individuals come from different cohorts, shared historical events are not so much of a confounding factor. By the same token, however, cross-sectional research may not be the most effective way to study differences between participants, as these differences may result not from their different ages but from their exposure to different historical events.

An accelerated longitudinal design or cross-sequential study combines both methodologies. Here, a researcher observes members of different birth cohorts at the same time, and then tracks all participants over time, charting changes in the groups. By comparing differences and similarities in development, one can more easily determine what changes can be attributed to individual or historical environment, and which are truly universal. Clearly such a study can be even more resource-consuming than a longitudinal study.

Additionally, these are all correlational, not experimental, designs, and so one cannot readily infer causation from the data they yield. Nonetheless, correlational research methods are common in the study of development, in part due to ethical concerns. In a study of the effects of poverty on development, for instance, one cannot easilly randomly assign certain families to a poverty condition and others to an affluent one, and so observation alone has to suffice.

EVOLUTIONARY PSYCHOLOGY

Evolutionary psychology (abbreviated ev-psych or EP) is a theoretical approach to psychology that attempts to explain "useful" mental and psychological traits-such as memory, perception, or language-as adaptations, i.e., as the functional products of natural selection. The purpose of this approach is to bring the functional way of thinking about biological mechanisms such as the immune system into the field of psychology, and to approach psychological mechanisms in a similar way. In short, evolutionary psychology is focused on how evolution has shaped the mind. Though applicable to any organism with a nervous system, most research in evolutionary psychology focuses on humans.

OVERVIEW

Evolutionary Psychology is, to quote Steven Pinker, "not a single theory but a large set of hypotheses" and a term which "has also come to refer to a particular way of applying evolutionary theory to the mind, with an emphasis on adaptation, gene-level selection, and modularity." Evolutionary Psychology proposes that the human brain comprises many functional mechanisms, called psychological adaptations or evolved cognitive mechanisms designed by the process of natural selection. Examples include language acquisition modules, incest avoidance mechanisms, cheater detection mechanisms, intelligence and sex-specific mating preferences, foraging mechanisms, alliance-tracking mechanisms, agent detection mechanisms, and so on. Evolutionary psychology has roots in cognitive psychology and evolutionary biology. It also draws on behavioral ecology, artificial intelligence, genetics, ethology, anthropology, archaeology, biology, and zoology. Evolutionary psychology is closely linked to sociobiology, but there are key differences between them including the emphasis on domain-specific rather than domain-general mechanisms, the relevance of measures of current fitness, the importance of mismatch theory, and psychology rather than behaviour. Many evolutionary psychologists, however, argue that the mind consists of both domain-specific and domain-general mechanisms, especially evolutionary developmental psychologists. Most sociobiological research is now conducted in the field of behavioral ecology.

The term evolutionary psychology was probably coined by Michael Ghiselin in his 1973 article in Science. Jerome Barkow, Leda Cosmides and John Tooby popularized the term in their highly influential 1992 book The Adapted Mind: Evolutionary Psychology and The Generation of Culture. Evolutionary psychology has been applied to the study of many fields, including economics, aggression, law, psychiatry, politics, literature, and sex.

General Evolutionary Theory

The idea that organisms are composed of a number of parts that serve different functions (i.e., living things are, in some sense, machines) goes back at least to Aristotle. This idea is the foundation of modern medicine and biology. William Paley, drawing upon the work of many others, argued that organisms are machines designed to function in particular environments. Paley believed that this evidence of 'design' was evidence for a designer-God. Charles Darwin appears to have been impressed with Paley's argument that organisms are designed for particular environments. The theory of natural selection, created by Charles Darwin and Alfred Russel Wallace, provided a scientific account of the origins of functional design in the natural world that did not invoke a supernatural designer.

Evolutionary psychology is ultimately rooted in the basic theoretical principles of evolutionary theory. It is sometimes seen not simply as a sub-discipline of psychology but as a way in which evolutionary theory can be used as a meta-theoretical framework within which to examine the entire field of psychology.

Natural selection, a key component of evolutionary theory, involves three main ingredients:

- Variation refers to a state in which there exists a variety of genetically determined traits within a population.
- Heritability refers to genetically determined traits which vary in their expression within a population (hair color, intelligence, height, anxiety-level, etc.)
- Selection refers to those heritable traits that remain in, and spread through, a population because those traits increased the reproduction of the organism.

Many traits that are selected for can actually hinder survival of the organism while increasing its reproductive opportunities. Consider the classic example of the peacock's tail. It is metabolically costly, cumbersome, and essentially a "predator magnet." What the peacock's tail does do is attract mates. Thus, the type of selective process that is involved here is what Darwin called sexual selection." Sexual selection can be divided into two types:

- Intersexual selection, which refers to the traits that one sex generally prefers in the other sex, (e.g. the peacock's tail).
- Intrasexual competition, which refers to the competition among members of the same sex for mating access to the opposite sex, (e.g. two stags locking antlers).

Darwin and Wallace proposed that natural and sexual selection, and not a supernatural designer, explain why organisms comprise a number of functional mechanisms that often exhibit surprisingly complex evidence of design. This theory has two important implications for the nature of these mechanisms, including the mechanisms of the brain. First, all evolved mechanisms must serve some function that ultimately increased the reproduction of the organism. Second, the design of each mechanism will be best understood in relation to the environment in which it evolved.

Inclusive Fitness: Inclusive fitness theory, which was proposed by William D. Hamilton in 1964 as a revision to evolutionary theory, is basically a combination of natural selection, sexual selection, and kin selection. It refers to the sum of an individual's own reproductive success plus the effects the individual's actions have on the reproductive success of their genetic relatives. General evolutionary theory, in its modern form, is essentially inclusive fitness theory.

Inclusive fitness theory resolved the issue of how "altruism" evolved. The dominant, pre-Hamiltonian view was that altruism evolved via group selection: the notion that altruism evolved for the benefit of the group. The problem with this was that if one organism in a group incurred any fitness costs on itself for the benefit of others in the group, (i.e. acted "altruistically"), then that organism would reduce its own ability to survive and/ or reproduce, therefore reducing its chances of passing on its altruistic traits. Furthermore, the organism that benefited from that altruistic act and only acted on behalf of its own fitness would increase its own chance of survival and/or reproduction, thus increasing its chances of passing on its "selfish" traits. Inclusive fitness resolved "the problem of altruism" by demonstrating that altruism can evolve via kin selection as expressed in Hamilton's rule:

$$\text{cost} < \text{relatedness} \times \text{benefit}$$

In other words, altruism can evolve as long as the fitness cost of the altruistic act on the part of the actor is less than the degree of genetic relatedness of the recipient times the fitness benefit to that recipient. This perspective reflects what is referred to as the gene-centered view of evolution and demonstrates that group selection is a very weak selective force. However, in recent years group selection has been making a comeback, (albeit a controversial one), as multilevel selection, which posits that evolution can act on many levels of functional organization, (including the "group" level), and not just the "gene" level.

Middle-level Evolutionary Theories

Middle-level evolutionary theories are theories that encompass broad domains of functioning. They are compatible with general evolutionary theory but not derived from it. Furthermore, they are applicable across species. During the early 1970's, three very important middle-level evolutionary theories were contributed by Robert Trivers:

- The theory of reciprocal altruism explains how altruism can arise amongst non-kin, as long as there is a sufficient probability of the recipient of the altruistic act reciprocating at a later date. The possibility was also noted by Trivers, later coined 'indirect altruism' by Richard Alexander, that reciprocation could be provided by third parties, raising the issue of social reputation.

- Parental investment theory refers to the different levels of investment in offspring on the part of each sex. For example, females in any species are defined as the sex with the larger gamete. In humans, females produce approximately one large, metabolically costly egg per month, as opposed to the millions of relatively tiny and metabolically cheap sperm that are produced each day by males. Females are fertile for only a few days each month, while males are fertile every day of the month. Females also have a nine month gestation period, followed by a few years of lactation. Males' obligatory biological investment can be achieved with one copulatory act. Consequently, human females have a significantly higher obligatory investment in offspring than males do. (In some species, the opposite is true.) Because of this difference in parental investment between males and females, the sexes face different adaptive problems in the domains of mating and parenting. Therefore, it is predicted that the higher investing sex will be more selective in mating, and the lesser investing sex will be more competitive for access to mates. Thus, differences in behaviour between sexes is predicted to exist not because of maleness or femaleness per se, but because of different levels of parental investment.

- The theory of parent-offspring conflict rests on the fact that even though a parent and his/her offspring are 50% genetically related, they are also 50% genetically different. All things being equal, a parent would want to allocate their resources equally amongst their offspring, while each offspring may want a little more for themselves. Furthermore, an offspring may want a little more resources from the parent than the parent is willing to give. In essence, parent-offspring conflict refers to a conflict of adaptive interests between parent and offspring.

However, if all things are not equal, a parent may engage in discriminative investment towards one sex or the other, depending on the parent's condition. Recall that females are the heavier parental investors in our species. Because of that, females have a better chance of reproducing at least once in comparison to males. Thus, according to the Trivers-Willard hypothesis, parents in good condition are predicted to favor investment in sons, and parents in poor condition are predicted to favor investment in daughters.

Products of the Evolutionary Process

All biological traits are products of the evolutionary process. Adaptationist methods, including evolutionary psychology, attempt to distinguish between adaptations and traits produced as by-products of adaptation.

- Adaptations are specially designed, species-typical traits that, (in terms of fitness costs), are "cost-efficient" and function to solve problems related to one's inclusive fitness. An example would be the Giant Panda's so-called 'thumb'.
- Evolutionary by-products are traits that were not specially designed for an adaptive function, although they may also be species-typical and may also confer benefits on the organism. A "spandrel" is a term coined by S.J.Gould and R.Lewontin (1979a) for traits which confer no adaptive advantage to an organism, but are 'carried along' by an adaptive trait. Their example being the lengthening of a bone in the hind limb of the Giant Panda, as a result of the lengthening of the corresponding bone in the forelimb. The bone in the forelimb has adapted to perform a function similar to that of an opposable thumb. The change in the hind limb, arguably, confers no adaptive advantage, whereas the change in the forelimb certainly does. The important implication of this idea for Evolutionary Psychology being, that therefore not every trait can be accounted for in terms of adaptive advantage. For instance, Gould puts forward the hypothesis that language itself in humans came about as a spandrel: "Natural selection made the human

> brain big, but most of our mental properties and potentials may be spandrels-that is, nonadaptive side consequences of building a device with such structural complexity" (S.J.Gould. The Pleasures of Pluralism; p.11) But of course, once a trait confers an adaptive advantage, as arguably most of our "mental properties and potentials" do, it is no longer a spandrel, and thus opens the debate concerning the importance of the concept of spandrels.

Natural selection is not the only evolutionary process that can change gene frequencies and produce novel traits. Genetic drift, or noise, refers to random effects resulting from chance variation in the genes, environment, or development. An example would be dry earwax in humans. Unlike natural selection, genetic drift does not produce complex, functionally designed adaptations.

Evolved Psychological Mechanisms: the Core of Evolutionary Psychology: Evolutionary psychology is based on the belief that, just like hearts, lungs, livers, kidneys, and immune systems, cognition has functional structure that has a genetic basis, and therefore has evolved by natural selection. Like other organs and tissues, this functional structure should be universally shared amongst a species, and should solve important problems of survival and reproduction. Evolutionary psychologists seek to understand psychological mechanisms by understanding the survival and reproductive functions they might have served over the course of evolutionary history.

Evolutionary psychologists subdivide the concept of psychological mechanisms into two general categories:

- Domain-specific mechanisms, which deal with recurrent adaptive problems over the course of human evolutionary history
- Domain-general mechanisms, which deal with evolutionary novelty

The Environment of Evolutionary Adaptedness: The term environment of evolutionary adaptedness, often abbreviated EEA, was coined by John Bowlby as part of attachment theory. It refers to the environment to which a particular evolved mechanism is adapted. More specifically, the EEA is defined as the set of historically recurring selection pressures that formed a given adaptation, as well as those aspects of the environment that were necessary for the proper development and functioning of the adaptation. In the environment in which ducks evolved, for example, attachment of ducklings to their mother had great survival value for the ducklings. Because the first moving being that a duckling was likely to see was its mother, a psychological mechanism that evolved to form an attachment to the first moving being would therefore properly function to form an attachment to the mother. In novel environments, however, the mechanism can malfunction by forming an attachment to a dog or human instead.

The genus Homo, which includes modern humans, appeared between 1.5 and 2.5 million years ago, a time that roughly coincides with the start of the Pleistocene 1.8 million years ago. Because the Pleistocene ended a mere 12,000 years ago, most human adaptations either newly evolved during the Pleistocene, or were maintained by stabilizing selection during the Pleistocene. Evolutionary psychology therefore proposes that the majority of human psychological mechanisms are adapted to reproductive problems frequently encountered in Pleistocene environments. In broad terms, these problems include those of growth, development, differentiation, maintenance, mating, parenting, and social relationships. To properly understand human mating psychology, for example, it is essential to recognize that in the EEA (as now) women got pregnant and men did not.

If humans are mostly adapted to Pleistocene environments, then some psychological mechanisms should occasionally exhibit "mismatches" to the modern environment, similar to the attachment patterns of ducks. One example is the fact that although about 30,000 people are killed with guns in the US annually (CDC pdf), whereas spiders and snakes kill only a handful, people nonetheless learn to fear spiders and snakes about as easily as they do a pointed gun, and more easily than an unpointed gun, rabbits or flowers (Öhman and Mineka 2001). A potential explanation is that spiders and snakes were a threat to human ancestors throughout the Pleistocene, whereas guns, rabbits and flowers were not. There is thus a mismatch between our evolved fear learning psychology and the modern environment. In sum, evolutionary psychology argues that to properly understand the functions of the brain one must understand the properties of the environment in which the brain evolved.

Controversies

Animal behavior studies have long recognized the role of evolution; the application of evolutionary theory to human psychology, however, is controversial. There are many families of criticism of the idea.

How Knowable is the Early Environment (EEA)?: Some critics of evolutionary psychology claim that because little is known about the Pleistocene, the evolutionary context in which humans developed (including population size, structure, lifestyle, eating habits, habitat, and more), there is little basis on which evolutionary psychology may operate. Most EP research is thus confined to certainties about the past, such as pregnancies only occurring in women, and that humans lived in groups.

Many evolutionary psychologists argue that this criticism is based on a misunderstanding. Evolutionary psychologists argue that they use knowledge of the environment of evolutionary adaptedness to generate hypotheses regarding possible psychological adaptations, and subsequently, these hypotheses can be tested and evaluated against the empirical evidence in just the same way that any other hypothesis generated from any other theoretical perspective can be assessed.

Furthermore, evolutionary psychologists posit that there are many environmental features that we can be sure played a part in our species' evolutionary history. They argue that our hunter-gatherer ancestors most certainly dealt with predators and prey, food acquisition and sharing, mate choice, child rearing, interpersonal aggression, interpersonal assistance, diseases and a host of other fairly predictable challenges that constituted significant selection pressures.

There also exists debate within evolutionary psychology about the nature of the EEA. Many evolutionary psychologists contend that many aspects of the EEA were not as consistent as other evolutionary psychologists would argue. This argument is used to support the notion that the mind consists of not only domain-specific psychological mechanisms but of domain-general ones as well, that deal with environmental novelty.

Too many Alternative Hypotheses: Critics claim that many valid hypotheses, including contradictory ones, can be drawn from the same evolutionary principles. Evolutionary psychology can predict many, or even all, behaviours for a given situation, including contradictory ones. Therefore many human behaviours will always fit some hypotheses. The central paradigm of evolutionary psychology is impossible to prove, and the predicting power of evolutionary psychology is doubtful. For example, kin selection predicts that humans will be altruistic toward relatives in proportion to their relatedness, while reciprocal altruism predicts that we will be altruistic toward people from whom we can expect altruism in the future (not strangers). However, altruism towards a complete stranger fits the handicap principle. Thus altruism toward every class of person fits some well-known hypothesis. This doesn't seem to give us any way to predict how a human will act in a given situation.

Evolutionary psychologists respond that their discipline is not primarily concerned with explaining the behavior of specific individuals, but rather broad categories of human behaviors across societies and cultures. It is the search for species-wide trends in behavior that distinguishes evolutionary psychology from cultural or social explanations. Thus hypotheses will attempt to explain contradictory human behavior because human behavior is often contradictory.

Falsifiability: Critics claim that many of the propositions of evolutionary psychology are not falsifiable. It is not possible to conduct definitive experiments on humans on an evolutionary timescale. This criticism in particular is common and often refers to evolutionary psychological theories as just-so stories.

The methods that evolutionary psychologists actually use for testing include comparing different species, comparing males and females, comparing individuals within a species and comparing the same individuals in different contexts. Their sources of data for testing evolutionary hypotheses include archeological records, data from hunter-gatherer societies, observations, self-reports, life-history data and public records, and human products. Source: Evolutionary Psychology: The New Science of the Mind by David Buss.

Biology versus Environment: Some studies have been criticized for their tendency to attribute to evolutionary processes elements of human cognition that may be attributable to social processes (e.g. preference for particular physical features in mates).

Evolutionary psychologists respond that many traits have been shown to be universal in humans and that social processes are related to evolutionary processes. They argue that statements such as "biology vs. environment" and "genes vs. culture" amount to false dichotomies. Evolutionary psychologists justify this claim by arguing that behavior results from an organism interacting with its environment. Psychological mechanisms, they argue, are created by genes, (which, in turn, were selected for by the evolutionary process), and those mechanisms help the organism negotiate its environment. Furthermore, they assert that many aspects of the environment, (e.g. culture and social institutions), are rested upon those mechanisms. In short, evolutionary psychologists argue that there is a bidirectional influence between things like "biology and environment" and "cognition and social processes."

Reductionism: Evolutionary Psychology is grounded on the theory that fundamentally our psychology is based on biology, the composition of our brains. This is a form of reductionism, a research philosophy according to which the nature of complex things can be understood in terms of simpler or more fundamental things (i.e. reduced). Importantly, reductionism as applied to consciousness and the brain, comes in various forms, some of which may be more philosophically problematic for Evolutionary Psychology than others. It is a long-standing matter of debate in Philosophy of Mind as to whether Mental States are identical with physical Brain States, or what sort of relation could be said to hold between the two.

However, it can be argued that Evolutionary Psychology is not committed to any particular answer to this question, being potentially compatible with Supervenience, Interactionism, Epiphenomenalism and both Token and Type physicalism, for example. Nevertheless, some people remain sceptical about any form of reduction from psychology to biology, and question its validity as an approach.

Why then, is the accusation of reductionism commonly used as a criticism of Evolutionary Psychology, and not say, as a criticism of Neuropsychopharmacology? (A discipline in which an implicit premise as regards psychology is that all states of mind, including both normal and drug-induced altered states, and diseases involving mental or cognitive dysfunction, have a neuro-chemical basis at the fundamental level.) The reason is that Evolutionary Psychology is a controversial field in itself, and to quote Richard Dawkins: "'reductionism', like sin, is one of those things only mentioned by people who are against it." The Blind Watchmaker, 1986 p.13. Here Dawkins makes a distinction between "direct" and "hierarchical" reductionism: organisms can be described in terms of DNA, DNA in terms of atoms, atoms in terms of sub-atomic particles etc.; but knowledge of sub-atomic particles will not directly explain animal or human behavior, nevertheless, one can make adequate explanations and predictions at a higher levels.

Issues in Ethics: That human psychology may be determined by our biology, which is shaped by our evolutionary past, is an important idea for those involved in Ethics. The implications are as broad and varied as the field of Ethics itself however, but generally, it is thought that Evolutionary Psychology describes factors which limit our free will, in that it can be seen to imply that we behave in ways in which we are 'naturally inclined'. Evolutionary Psychology is not in itself of course a theory of Ethics, it merely states what is, not what ought to be, but this idea of 'natural inclination' is a recurring theme in Ethics, arising not just out of Evolutionary Psychology, but almost all of the Social Sciences.

To quote J. Mizzoni from his article Ruse's Darwinian Ethics and Moral Realism: "There are some moral philosophers (e.g., Thomas Nagel) who believe that evolutionary considerations are irrelevant to a full understanding of the foundations of ethics. Other moral philosophers (e.g., J.L. Mackie) tell quite a different story. They hold that the admission of the evolutionary origins of human beings compels us to concede that there are no foundations for ethics." As we can see, the matter is subject to debate, some believing that evolutionary considerations are correct and necessary, others that they are incorrect and a form of the naturalistic fallacy, and others indeed that they are morally harmful whether correct or not. This idea of the supposed moral harm done by bringing together the Social Sciences and Ethics is arguably where the greatest controversy lies.

Empirical Evidence: Some commentators, like philosopher David Buller, agree with the general argument that the human mind has evolved over time but disagree with the specific claims evolutionary psychologists make. Buller has argued, among other things, that the so-called Cinderella Effect, the argument that there are gender differences with respect to sexual jealousy, and the contention that the mind consists of thousands of modules, are unsupported by the available empirical evidence.

An alternative to the "mental module" view of how human minds evolved is offered by cognitive psychologist Merlin Donald. He argues that over evolutionary time the mind has gained adaptive advantage from a general problem solver. Donald articulates this view in his book "A Mind So Rare: The Evolution of Human Consciousness".

Another criticism regarding empirical evidence is a perceived lack of cross-cultural proofs. Many empirical studies come from a single culture. In these cases it is unknown if discoveries hold across all cultures.

EXPERIMENTAL PSYCHOLOGY

Experimental psychology is an approach to psychology that treats it as one of the natural sciences, and therefore assumes that it is susceptible to the experimental method. Many experimental psychologists have gone further, and have assumed that all methods of investigation other than experimentation are suspect. In particular, experimental psychologists have been inclined to discount the case study and interview methods as they have been used in clinical and developmental psychology.

Since it is a methodological rather than a substantive category, experimental psychology embraces a disparate collection of areas of study. It is usually taken to include the study of perception, cognitive psychology, comparative psychology, the experimental analysis of behavior, and some aspects of physiological psychology and developmental psychology.

Wilhelm Wundt, the founder of structuralism school of psychology is considered as the father of experimental psychology. He established the first psychology laboratory in Leipzig. The earliest experimental psychologists, including Hermann Ebbinghaus and Edward Titchener included introspection among their experimental methods. However, in the first half of the twentieth century, experimental psychology became closely allied with behaviourism, especially in the United States, and this led to some neglect of mental phenomena. In Europe including the United Kingdom this was less so, and under the influence of psychologists such as Sir Frederic Bartlett, Kenneth Craik, W. E. Hick and Donald Broadbent, experimental psychologists focused on topics such as thinking, memory and attention, laying the foundations for the subsequent development of cognitive psychology. Wundt was the first to call himself a "psychologist". He is also the first research/ experimental psychologist.

With the expansion of psychology as a discipline in the later half of the twentieth century, and the growth in the size and number of its subfields, the phrase "experimental psychology" has come to cover many areas. Many psychologists would now identify with a smaller field such as cognitive or comparative psychology. Many experimental psychologists use a range of methods and do not confine themselves to a strictly experimental approach. Developments in the philosophy of science have also had an impact on the exclusive prestige of experimentation. Conversely, the experimental method is now widely used in fields such as developmental and social psychology that were not part of experimental psychology as it was originally conceived. The phrase continues in use, however, in the titles of a number of well established, high prestige learned societies and scientific journals, as well as some university courses of study in psychology.

PSYCHOLINGUISTICS

Psycholinguistics or psychology of language is the study of the psychological and neurobiological factors that enable humans to acquire, use, and understand language. Initial forays into psycholinguistics were largely philosophical ventures, due mainly to a lack of cohesive data on how the human brain functioned. Modern research makes use of biology, neuroscience, cognitive science, and information theory to study how the brain processes language. There are a number of subdisciplines; for example, as non-invasive techniques for studying the neurological workings of the brain become more and more widespread, neurolinguistics has become a field in its own right.

Psycholinguistics covers the cognitive processes that make it possible to generate a grammatical and meaningful sentence out of vocabulary and grammatical structures, as well as the processes that make it possible to understand utterances, words, text, etc. Developmental psycholinguistics studies infants' and children's ability to learn language, usually with experimental or at least quantitative methods (as opposed to naturalistic observations such as those made by Jean Piaget in his research on the development of children).

Areas of Study: Psycholinguistics is interdisciplinary in nature and is studied by people in a variety of fields, such as psychology, cognitive science, and linguistics. There are several subdivisions within psycholinguistics that are based on the components that make up human language.

- Phonetics and phonology are concerned with the study of speech sounds. Within psycholinguistics, research focuses on how the brain processes and understands these sounds.
- Morphology is the study of word structures, especially the relationships between related words (such as dog and dogs) and the formation of words based on rules (such as plural formation).

- Syntax is the study of the patterns which dictate how words are combined together to form sentences.
- Semantics deals with the meaning of words and sentences. Where syntax is concerned with the formal structure of sentences, semantics deals with the actual meaning of sentences.
- Pragmatics is concerned with the role of context in the interpretation of meaning.
- The study of word recognition and reading examines the processes involved in the extraction of orthographic, morphological, phonological, and semantic information from patterns in printed text

Theories: Theories about how language works in the human mind attempt to account for, among other things, how we associate meaning with the sounds (or signs) of language and how we use syntax-that is, how we manage to put words in the proper order to produce and understand the strings of words we call "sentences." The first of these items-associating sound with meaning-is the least controversial and is generally held to be an area in which animal and human communication have at least some things in common. Syntax, on the other hand, is controversial, and is the focus of the discussion that follows.

There are essentially two schools of thought as to how we manage to create syntactic sentences: (1) syntax is an evolutionary product of increased human intelligence over time and social factors that encouraged the development of spoken language; (2) language exists because humans possess an innate ability, an access to what has been called a "universal grammar." This view holds that the human ability for syntax is "hard-wired" in the brain. This view claims, for example, that complex syntactic features such as recursion are beyond even the potential abilities of the most intelligent and social non-humans. (Recursion, for example, includes the use of relative pronouns to refer back to earlier parts of a sentence-"The girl whose car is blocking my view of the tree that I planted last year is my friend.") The ability to use syntax like that would not exist without an innate concept that contains the underpinnings for the grammatical rules that produce recursion, says the "innate" view. Children acquiring a language, thus, have a vast search space to explore among possible human grammars, settling, logically, on the language(s) spoken or signed in their own community of speakers. Such syntax is, according to the second point of view, what defines human language and makes it different from even the most sophisticated forms of animal communication.

The first view was prevalent until about 1960 and is well represented by the mentalistic theories of Jean Piaget and the empiricist, Rudolf Carnap. As well, the school of psychology known as behaviorism puts forth the point of view that language-syntax included-is behavior shaped by conditioned response. The second point of view-the "innate" one-can fairly be said to have begun with Noam Chomsky's highly critical review of Skinner's book in 1959 in the pages of the journal Language. That review started what has been termed "the cognitive revolution" in psychology.

The field of psycholinguistics since then has been defined by reactions to Chomsky, pro and con. The pro view still holds that the human ability to use syntax is qualitatively different from any sort of animal communication. That ability might have resulted from a favorable mutation (extremely unlikely) or (more likely) from an adaptation of skills evolved for other purposes. That is, precise syntax might, indeed, serve group needs; better linguistic expression might produce more cohesion, cooperation, and potential for survival, BUT precise syntax can only have developed from rudimentary-or no-syntax, which would have had no survival value and, thus, would not have evolved at all. Thus, one looks for other skills, the characteristics of which might have later been useful for syntax. In the terminology of modern evolutionary biology, these skills would be said to be "pre-adapted" for syntax. Just what those skills might have been is the focus of recent research-or, at least, speculation.

The con view still holds that language-including syntax-is an outgrowth of hundreds of thousands of years of increasing intelligence and tens of thousands of years of human interaction. From that view, syntax in language gradually increased group cohesion and potential for survival. Language-syntax and all-is a cultural artifact. This view challenges the "innate" view as scientifically unfalsifiable; that is to say, it can't be tested; the fact that a particular, conceivable syntactic structure does not exist in any of the world's finite repertoire of languages is an interesting observation, but it is not proof of a genetic constraint on possible forms, nor does it prove that such forms couldn't exist or couldn't be learned.

Contemporary theorists, besides Chomsky, working in the field of theories of psycholinguistics include George Lakoff, Steven Pinker, and Michael Tomasello.

Methodologies: Much methodology in psycholinguistics takes the form of behavioral experiments. In these types of studies, subjects are presented with some form of linguistic input and asked to perform a task (e.g. make a judgement, reproduce the stimulus, read a visually presented word aloud). Reaction times (usually on the order of milliseconds) and proportion of correct responses are the most often employed measures of performance.

Such tasks might include, for example, asking the subject to convert nouns into verbs; e.g., "book" suggests "to write," "water" suggests "to drink," and so on. Another experiment might present an active sentence such as "Bob threw the ball to Bill" and a passive equivalent, "The ball was thrown to Bill by Bob" and then ask the question, "Who threw the ball?" We might then conclude (as is the case) that active sentences are processed more easily (faster) than passive sentences. More interestingly, we might also find out (as is the case) that some people are unable to understand passive sentences; we might then make some tentative steps towards understanding certain types of language deficits (generally grouped under the broad term, aphasia).

Until the recent advent of non-invasive medical techniques, brain surgery was the preferred way for language researchers to discover how language works in the brain. For example, severing the corpus callosum (the bundle of nerves that connects the two hemispheres of the brain) was at one time a treatment for some forms of epilepsy. Researchers could then study the ways in which the comprehension and production of language were affected by such drastic surgery. Where an illness made brain surgery necessary, language researchers had an opportunity to pursue their research.

Newer, non-invasive techniques now include brain imaging by positron emission tomography (PET); functional magnetic resonance imaging (fMRI); event related potentials (ERP) and transcranial magnetic stimulation (TMS). Brain imaging techniques vary in their spatial and temporal resolutions (fMRI has a resolution of a few thousand neurons per pixel, and ERP has millisecond accuracy). Each type of methodology presents a set of advantages and disadvantages for studying a particular problem in psycholinguistics.

Computational modelling-e.g. the DRC model of reading and word recognition proposed by Coltheart and colleagues-is another methodology. It refers to the practice of setting up cognitive models in the form of executable computer programs. Such programs are useful because they require theorists to be explicit in their hypotheses and because they can be used to generate accurate predictions for theoretical models that are so complex that they render discursive analysis unreliable.

More recently, eye tracking has been used to study online language processing. Beginning with Rayner (1978) the importance and informativity of eye-movements during reading was established. Tanenhaus et al., have performed a number of visual-world eye-tracking studies to study the cognitive processes related to spoken language. Since eye movements are closely linked to the current focus of attention, language processing can be studied by monitoring eye movements while a subject is presented with linguistic input.

Issues and Areas of Research

Developmental: There are a number of unanswered questions in psycholinguistics. In part, they are suggested by some of the items mentioned in the section on "theories" (above). For example, is the human ability to use syntax based on innate mental structures or is syntactic speech the function of intelligence and interaction with other humans? Can we even design psycholinguistic experiments to find that out? Research in animal communication has much to offer here. Can some animals be taught the syntax of human language? If so, what does that mean? If not, what does that mean?

How are infants able to learn language? Almost all healthy human infants acquire language readily in the first few years of life. This is true across cultures and societies. And what about children who do not learn language properly? There is a broad field called aphasia that deals with language deficits. Can research in psycholinguistics ever be of some therapeutic value? In addition, it is much more difficult for adults to acquire second languages than it is for infants to learn their first language (bilingual infants are able to learn both of their native languages easily). Thus, critical periods may exist during which language is able to be learned readily. A great deal of research in psycholinguistics focuses on how this ability develops and diminishes over time. It also seems to be the case that the more languages one knows, the easier it is to learn more.

Also, recent research using new non-invasive imaging techniques seeks to shed light on just where language is located in the brain. How localized is language? How distributed is it from one hemisphere to the other? The older, traditional descriptions of the language functions of Broca's area, Wernicke's area and other areas of the brain will be refined as research continues.

PSYCHOPHYSICS

Psychophysics is a subdiscipline of psychology dealing with the relationship between physical stimuli and their subjective correlates, or percepts.

Gustav Theodor Fechner founded psychophysics in 1860 with the publication of Elemente der Psychophysik. He described research relating physical stimuli with how they are perceived and set out the philosophical foundations of the field. Fechner wanted to develop a theory that could relate matter to the mind, by describing the relationship between the world and the way it is perceived (Snodgrass, 1975). Fechner's work formed the basis of psychology as a science. Wilhelm Wundt, the founder of the first laboratory for psychological research, built upon Fechner's work.

Psychophysicists usually employ experimental stimuli that can be objectively measured, such as pure tones varying in intensity, or lights varying in luminance. All the senses have been studied: vision, hearing, touch (including-skin and enteric perception), taste, smell, and the sense of time. Regardless of the sensory domain, there are three main topics in the psychophysical classification scheme: absolute thresholds, discrimination thresholds, and scaling.

The most common use of psychophysics is in producing scales of human experience of various aspects of physical stimuli. Take for an example the physical stimulus of frequency of sound. Frequency of a sound is measured in hertz, cycles per second. But human experience of the frequencies of sound is not the same as the frequencies. For one thing, there is a frequency below which no sounds can be heard, no matter how intense they are (around 20 Hz depending on the individual) and there is a frequency above which no sounds can be heard, no matter how intense they are (around 20,000 Hz, again depending on the individual). For another, doubling the frequency of a sound (e.g., from 100 Hz to 200 Hz) does not lead to a doubling of experience. The perceptual experience of the frequency of sound is called pitch, and it is measured by psychophysicists in mels.

More analytical approaches allow the use of psychophysical methods to study neurophysiological properties and sensory processing mechanisms. This is of particular importance in human research, where other (more invasive) methods are not used due to ethical reasons.

Areas of investigation include sensory thresholds, methods of measurement of sensitivity, and signal detection theory.

Experimentation: In psychophysics, experiments seek to determine whether the subject can detect a stimulus, identify it, differentiate between it and another stimulus, and describe the magnitude or nature of this difference (Snodgrass, 1975).

Thresholds: A threshold (or limen), is the point of intensity at which the participant can just detect the presence of, or difference in, a stimulus. Stimuli with intensities below the threshold are considered not detectable, however stimuli at values close to threshold will often be detectable some proportion of the time. Due to this, a threshold is considered to be the point at which a stimulus, or change in a stimulus, is detected some proportion of the time (Snodgrass, 1975).

There are two kinds of thresholds: absolute and difference. An absolute threshold is the level of intensity of a stimulus at which the subject is able to detect the presence of the stimulus some proportion of the time (a p level of 50% is often used). An example of an absolute threshold is the number of hairs on the back of one's hand that must be touched before it can be felt-a participant may be unable to feel a single hair being touched, but may be able to feel two or three as this exceeds the threshold.

A difference threshold is the magnitude of the difference between two stimuli of differing intensities that the participant is able to detect some proportion of the time (again, 50% is often used). To test this threshold, several difference methods are used. The subject may be asked to adjust one stimulus until it is perceived as the same as the other, may be asked to describe the magnitude of the difference between two stimuli, or may be asked to detect a stimulus against a background.

Absolute and difference thresholds are sometimes considered similar because there is always background noise interfering with our ability to detect stimuli (Snodgrass, 1975), however study of difference thresholds still occurs, for example in pitch discrimination tasks.

Classic Methods of Experimentation: Psychophysical experiments have traditionally used three methods for testing subjects' perception in stimulus detection and difference detection experiments: the method of limits, the method of constant stimuli, and the method of adjustment (Snodgrass, 1975).

Method of Limits: Wilhelm Wundt invented the method of limits. The subject reports whether he or she detects the stimulus. In ascending method of limits, some property of the stimulus starts out at a level so low that the stimulus could not be detected, then this level is gradually increased until the participant reports that they are aware of it. For example, if the experiment is testing the minimum amplitude of sound that can be detected, the sound begins too quietly to be perceived, and is made gradually louder. In the descending method of limits, this is reversed. In each case, the threshold is considered to be the level of the stimulus property at which the stimuli is just detected.

In experiments, the ascending and descending methods are used alternately and the thresholds are averaged. A possible disadvantage of these methods is that the subject may become accustomed to reporting that they perceive a stimulus and may not continue reporting the same way even beyond the threshold (the error of habituation). Conversely, the subject may also anticipate that the stimulus is about to become detectable or undetectable and may make a premature judgment (the error of expectation). To avoid these potential pitfalls, Georg von Bekesy introduced the staircase method in 1960 in his study of auditory perception. In this method, the sound starts out audible and gets quieter after each of the subject's responses, until the subject does not report hearing it. At that point, the sound is made louder at each step, until the subject reports hearing it, at which point it is made quieter in steps again. This way the experimenter is able to "zero in" on the threshold.

Method of Constant Stimuli: Instead of being presented in ascending or descending order, in the method of constant stimuli the levels of a certain property of the stimulus are not related from one trial to the next, but presented randomly. This prevents the subject from being able to predict the level of the next stimulus, and therefore reduces errors of habituation and expectation. The subject again reports whether he or she is able to detect the stimulus.

Method of Adjustment: Also called the method of average error, the method of adjustment asks the subject to control the level of the stimulus, instructs them to alter it until it is just barely detectable against the background noise, or is the same as the level of another stimulus.

Non-traditional Methods of Experimentation

Staircase Procedure: Often, the classic methods of experimentation are argued to be inefficient. This is because, in advance of testing, the psychometric threshold is usually unknown and a lot of data has to be collected at points on the psychometric function that provide little information about its shape (the tails).

Adaptive staircase procedures can be used such that the points sampled are clustered around the psychometric threshold. However, the cost of this efficiency, is that you do not get the same amount of information regarding the shape of the psychometric function as you can through classical methods. Despite this, it is still possible to estimate the threshold and slope by fitting psychometric functions to the obtained data, although estimates of psychometric slope are likely to be more variable than those from the method of constant stimuli (for a reasonable sampling of the psychometric function).

Discrimination: In discrimination experiments, the experimenter seeks to determine at what point the difference between two stimuli, such as two weights or two sounds, is detectable. The subject is presented with one stimulus, for example a weight, and is asked to say whether another weight is heavier or lighter (in some experiments, the subject may also say the two weights are the same). At the point of subjective equality (PSE), the subject perceives the two weights to be the same. The just noticeable difference (JND), or difference limen (DL), is the difference in stimuli that the subject notices some proportion of the time (50% is usually used for p).

The methods of limits, constant stimuli and adjustment can be used in difference detection by asking the subject to detect a difference between stimuli rather than detect a single stimulus.

PERSONALITY PSYCHOLOGY

Personality psychology is a branch of psychology which studies personality and individual differences. One emphasis in this area is to construct a coherent picture of a person and his or her major psychological processes. Another emphasis views personality as the study of individual differences, in other words, how people differ from each other. A third area of emphasis examines human nature and how all people are similar to one other. These three viewpoints merge together in the study of personality.

Personality can be defined as a dynamic and organized set of characteristics possessed by a person that uniquely influences his or her cognitions, motivations, and behaviors in various situations (Ryckman, 2004). The word "personality" originates from the Latin persona, which means mask. Significantly, in the theatre of the ancient Latin-speaking world, the mask was not used as a plot device to disguise the identity of a character, but rather was a convention employed to represent or typify that character.

The pioneering American psychologist, Gordon Allport (1937) described two major ways to study personality, the idiographic and the nomothetic. Nomothetic psychology seeks general laws that can be applied to many different people, such as the principle of self-actualization, or the trait of extraversion. Idiographic psychology is an attempt to understand the unique aspects of a particular individual.

The study of personality has a rich and varied history in psychology, with an abundance of theoretical traditions. Some psychologists have taken a highly scientific approach, whereas others have focused their attention on theory development. There is also a substantial emphasis on the applied field of personality testing.

Philosophical Assumptions

Many of the ideas developed by historical and modern Personality Theorists stem from basic philosophical assumptions they hold. Psychology is not a purely empirical discipline, as it brings in elements of art, science, and philosophy to draw general conclusions. The following five categories are some of the most fundamental philosophical assumptions where theorists disagree:

Freedom versus Determinism: The debate over whether we have control over our own behavior and understand the motives behind it (Freedom), or if our behavior is basically determined by some other force over which we might not have control (Determinism). We may merely respond to external forces like government, parents, professors, the economic system, etc.; or we may even be constrained to behave in certain ways by our genetics, upbringing, etc.

Heredity versus Environment: The nature vs. nurture debate. Is personality determined largely by genetics and heredity, or by environment and experiences? There is evidence for both sides, and the field of behavior genetics has demonstrated interesting new results in this area, such as the importance of unshared environmental effects (such as individual experiences) over shared environmental effects (such as family and demographics). Furthermore, twin studies indicate that there is a substantial genetic contribution to most personality traits.

Uniqueness versus Universality: The argument over whether we are all unique individuals (Uniqueness) or if humans are basically similar in their nature (Universality).

Proactive versus Reactive: Do we primarily act through our own initiative (Proactive), or do we react to outside stimuli (Reactive)?

Optimistic versus Pessimistic: Finally, whether or not we can alter our personalities (Optimistic) or if they remain the same throughout our whole lives (Pessimistic).

PERSONALITY THEORIES

There are several theoretical perspectives on personality in psychology, which involve different ideas about the relationship between personality and other psychological constructs, as well as different theories about the way personality develops.

Critics of personality theory claim that personality is "plastic" across time, places, moods, and situations. Changes in personality may indeed result from diet (or lack thereof), medical effects, significant events, or learning. However, most personality theories emphasize stability over fluctuation.

Trait Theories: According to the Diagnostic and Statistical Manual of the American Psychiatric Association, personality traits are "enduring patterns of perceiving, relating to, and thinking about the environment and oneself that are exhibited in a wide range of social and personal contexts." Theorists generally assume that (a) traits are relatively stable over time, (b) traits differ among individuals (e.g. some people are outgoing while others are shy), and c) traits influence behavior.

The most common models of traits incorporate three to five broad dimensions or factors. The least controversial dimension, observed as far back as the ancient Greeks, is simply extraversion vs. introversion (outgoing and physical-stimulation-oriented vs. quiet and physical-stimulation-averse).

- Gordon Allport delineated different kinds of traits, which he also called dispositions. Central traits are basic to an individual's personality, while secondary traits are more peripheral. Common traits are those recognized within a culture and thus may vary from culture to culture. Cardinal traits are those by which an individual may be strongly recognized.
- Raymond Cattell's research propagated a two-tiered personality structure with sixteen "primary factors" (16 Personality Factors) and five "secondary factors." A different model was proposed by Hans Eysenck, who believed that just three traits-extraversion, neuroticism and psychoticism-were sufficient to describe human personality. Differences between Cattell and Eysenck emerged due to preferences for different forms of factor analysis, with Cattell using oblique, Eysenck orthogonal, rotation to analyse the factors that emerged when personality questionnaires were subjected to statistical analysis. Today, the Big Five factors have the weight of a considerable amount of empirical research behind them. Building on the work of Cattell and others, Lewis Goldberg proposed a five-dimension personality model, nicknamed the "Big Five":
  1. Extraversion-outgoing and stimulation-oriented vs. quiet and stimulation-avoiding
  2. Neuroticism-emotionally reactive, prone to negative emotions vs. calm, imperturbable, optimistic
  3. Agreeableness-affable, friendly, conciliatory vs. aggressive, dominant, disagreeable
  4. Conscientiousness-dutiful, planful, and orderly vs. laidback, spontaneous, and unreliable
  5. Openness to experience-open to new ideas and change vs. traditional and oriented toward routine

- John L. Holland's RIASEC vocational model, commonly referred to as the Holland Codes, stipulates that there are six personality traits that lead people to choose their career paths. This model is widely used in vocational counseling and is a circumplex model where the six types are represented as a hexagon where adjacent types are more closely related than those more distant.

Trait models have been criticized as being purely descriptive and offering little explanation of the underlying causes of personality. Eysenck's theory, however, does propose biological mechanisms as driving traits, and modern behavior genetics researchers have demonstrated a clear genetic substrate to them. Another potential weakness with trait theories is that they lead people to accept oversimplified classifications, or worse offer advice, based on a superficial analysis of one's personality. Finally, trait models often underestimate the effect of specific situations on people's behavior. It is important to remember that traits are statistical generalizations that do not always correspond to an individual's behavior.

Type Theories: Personality type refers to the psychological classification of different types of people. Personality types are distinguished from personality traits, which come in different levels or degrees. According to type theories, for example, there are two types of people, introverts and extraverts. According to trait theories, introversion and extraversion are part of a continuous dimension, with many people in the middle. The idea of psychological types originated in the theoretical work of Carl Jung.

Building on the writings and observations of Carl Jung, during WWII Isabel Briggs Myers and her mother Katharine C. Briggs delineated personality types by constructing the Myers-Briggs Type Indicator. This model was later elaborated further by David Keirsey. It is an older, more theoretical approach to personality, accepting Extraversion vs. Introversion as basic and further adding three more dimensions:

1. Extraversion vs. Introversion
2. Intuition vs. Sensing (trust in conceptual/abstract models of reality versus concrete sensory-oriented facts)

3. Thinking vs. Feeling (thinking as the prime-mover in decision-making vs. feelings as the prime-mover in decision-making)
4. Perceiving vs. Judging (desire to perceive events vs. desire to have things done so judgements can be made)

This personality typology has some aspects of a trait theory: it explains people's behaviour in terms of opposite fixed characteristics. In these more traditional models, the intuition factor is considered the most basic, dividing people into "N" or "S" personality types. An "N" is further assumed to be guided by the thinking or objectication habit, or feelings, and be divided into "NT" (scientist, engineer) or "NF" (author, human-oriented leader) personality. An "S", by contrast, is assumed to be more guided by the perception axis, and thus divided into "SP" (performer, craftsman, artisan) and "SJ" (guardian, accountant, bureaucrat) personality.

These four are considered basic, with the other two factors in each case (including always extraversion) less important. Critics of this traditional view have observed that the types are quite strongly stereotyped by professions, and thus may arise more from the need to categorize people for purposes of guiding their career choice. This among other objections led to the emergence of the five factor view, which is less concerned with behavior under work stress and more concerned with behavior in personal and emotional circumstances. Some critics have argued for more or fewer dimensions while others have proposed entirely different theories (often assuming different definitions of "personality").

Type A personality: During the 1950s, Meyer Friedman and his co-workers defined what they called Type A and Type B behavior patterns. They theorized that intense, hard-driving Type A personalities had a higher risk of coronary disease because they are "stress junkies." Type B people, on the other hand, tended to be relaxed, less competitive, and lower in risk. There was also a Type AB mixed profile. Dr. Redford Williams, cardiologist at Duke University, refuted Friedman's theory that Type A personalities have a higher risk of coronary heart disease; however, current research indicates that the hostility component of Type A may have health implications.

Type A/B theory has been extensively criticized by psychologists because it tends to oversimplify the many dimensions of an individual's personality.

Psychoanalytic Theories: Psychoanalytic theories explain human behaviour in terms of the interaction of various components of personality. Sigmund Freud was the founder of this school. Freud drew on the physics of his day (thermodynamics) to coin the term psychodynamics. Based on the idea of converting heat into mechanical energy, he proposed that psychic energy could be converted into behavior. Freud's theory places central importance on dynamic, unconscious psychological conflicts.

Freud broke the human personality down to three significant components: the ego, superego, and id. The id is the source of sexual energy that builds up and needs to be released or expressed in some way. The id is motivated by the pleasure principle. The ego is the structure that helps the id express itself. It emerges in order to realistically meet the wishes and demands of the id in accordance with the outside world. It operates according to the reality principle. Finally, the superego exercises moral judgement and societal rules in keeping the ego and id in check. The superego is the last function of the personality to develop and may be seen as an outcome of the interactions with one's parents during the long period of childhood dependency. According to Freud, personality is based on the interaction of these three components.

Sexuality is a major component of Freud's theory. He believed that humans are sexual throughout childhood, though it is important to note that Freud's broad understanding of sexuality included all kinds of pleasurable feelings experienced by the human body. Freud proposed five psychosexual stages of personality development:

1. Oral Stage-birth to approximately age one
2. Anal Stage-two years of age
3. Phallic Stage-between three and six
4. Latency Period-about seven years old to puberty
5. Genital Stage-occurs during adolescence.

Freud believed that adult personality is determined by early childhood experiences. He suggested that events in the past could influence the present, such as when a person develops a fixation during one of these five stages and is unable to develop further.

One of Sigmund Freud's earlier associates Alfred Adler did agree with Freud that early childhood experiences are important to development, and believed that birth order may influence personality development. Adler believed the oldest was the one that set high goals to achieve to get attention back that they lost when the younger siblings were born. He believed the middle children were competitive and ambitious possibly so they are able to surpass the first-born's achievements, but were not as much concerned about the glory. Also he believed that the last born would be more dependent and sociable but be the baby. He also believed that only children love being the center of attention and mature quickly, but in the end fail to become independent.

Heinz Kohut thought similarly to Freud's idea of transference. He used narcissism as a model of how we develop our sense of self. Narcissism is the exaggerated sense of one self in which is believed to exist in order to protect one's low self esteem and sense of worthlessness. Kohut had a significant impact on the field by extending Freud's theory of narcissism and introducing what he called the 'self-object transferences' of mirroring and idealization. In other words, children need to idealize and emotionally "sink into" and identify with the idealized competence of admired figures such as parents or older siblings. They also need to have their self-worth mirrored by these people. These experiences allow them to thereby learn the self-soothing and other skills that are necessary for the development of a healthy sense of self.

Another important figure in the world of personality theory would be Karen Horney. She is credited with the development of the "real self" and the "ideal self". She believes that all people have these two views of their own self. The "real self" is how you really are with regards to personality, values, and morals; but the "ideal self" is a construct you apply to yourself to conform to social and personal norms and goals. Ideal self would be "I can be successful, I am CEO material"; and real self would be "I just work in the mail room, with not much chance of high promotion".

Margaret Mahler agreed with Klein's theory of linking relationships children have with their mothers to mental disorders of disturbed children. Certain disorders directly relate to what kind of relationship they had with their mothers. An example of this would be people diagnosed with schizophrenia. They are often too attached to their mother as children and even become obsessed, and never get over the "Oedipus" or "Electra" complex. Another example would be autistic children. Autistic children show no interest in their mother, relating to her, and so on. Both of these are very opposite reactions, but both have to do with the outcome of the mental disorder.

Freud used the term "object" to refer to any target that an infant uses to satisfy his or her needs. In Object Relations Theory the object is the aim of "relational needs" in human development. These objects are most often people, such as primary caretakers and significant others. However, in young children these objects may include a blanket, favorite toy, pacifiers, etc. The child becomes attached to the object because it provides pleasure for the child. When very young he is unable to distinguish himself from the object. For the child, the transitional object provides a connection between the child's inner and outer worlds. The child learns about separateness between subjective and objective. From birth through life, object relations theorists propose that individuals seek to develop human relationships and form attachments that may aid or hinder their development. (Engler, 2006).

Behaviorist Theories: Behaviorists explain personality in terms of reactions to external stimuli, and was a radical shift away from Freudian philosophy. This school of thought was developed by B. F. Skinner who put forth a model which emphasized the mutual interaction of the person 'or "the organism" with its environment. Skinner believed that children do bad things in order to get the attention that they crave. For example: a child cries because the child desires attention and knows it will be given. These are the stimulus, response, and consequences. The stimulus is the child being ignored, the response is the child acting out, and the attention that child gets is the consequence. According to this theory, people's behavior is formed by processes such as operant conditioning. Skinner put forward a 'three term contingency model' which helped promote analysis of behavior based on the 'Stimulus-Response-Consequence Model' in which the critical question is: "Under which circumstances or antecedent "stimuli" does the organism engage in a particular behavior or "response," which in turn produces a particular "consequence"?"

Ivan Pavlov is another notable influence. He is well known for his classical conditions experiments involving a dog. These physiological studies on this dog led him to discover the foundation of behaviorism as well as classical conditioning. Pavlov would begin his experiment by first ringing a bell, which would cause no response from the dog. He would proceed to place food in front of the dog's face, causing the dog to salivate. Several seconds later, he would ring the bell again, causing the dog to now salivate. After continuing this experiment several times, the dog would salivate at just the ring of the bell. These conditioning experiments can be used for many different types of experiments. He can do these experiments for any situation. For example, If every time someone ate fish they got sick, and he rang the bell when they ate the fish, eventually they would get sick from just the sound of the bell.

John B. Watson, The Father of American Behaviorism, made four major assumptions about radical Behaviorisms-

1. *Evolutionary Continuity:* The laws of behavior are applied equally to all living organisms, so we can study animals as simple models of complex human responses.
2. *Reductionism:* All behaviors are linked to physiology.
3. *Determinism:* Animals do not respond freely, they respond in a programmed way to external stimuli. Biological organisms respond to outside influences.
4. *Empiricism:* Only our actions are observable evidence of our personality. Psychology should involve the study of observable (overt) behavior.

All behaviorists focus on observable behavior. Thus there is no emphasis on unconscious motives, internal traits, introspection, or self analysis. Behavior modification is a form of therapy that applies the principles of learning to achieve changes in behavior.

Cognitive Theories: In cognitivism, behaviour is explained as guided by cognitions (e.g. expectations) about the world, especially those about other people. Cognitive theories are theories of personality that emphasize cognitive processes such as thinking and judging. Albert Bandura, a social learning theorist suggested that the forces of memory and emotions worked in conjunction with environmental influences. Bandura was known mostly for his studies involving his "bobo doll." During these experiments, Bandura video taped a college student kicking and verbally abusing a bobo doll. He then showed this video to a class of kindergartners who were getting ready to go out to play. When they entered the play room, they saw bobo dolls, and some hammers. The people observing these children at play saw a group of children beating the doll. He called this study and his findings observational learning, or modeling.

Early examples of approaches to cognitive style are listed by Baron (1982). These include Witkin's (1965) work on field dependency, Gardner's (1953) discovering people had consistent preference for the number of categories they used to categorise heterogeneous objects, and Block and Petersen's (1955) work on confidence in line discrimination judgments. More central to this field have been:

- Self-efficacy work, dealing with confidence people have in abilities to do tasks (Bandura, 1997);
- Locus of control theory (Lefcourt, 1966; Rotter, 1966) dealing with different beliefs people have about whether their worlds are controlled by themselves or external factors;
- Attributional style theory (Abramson, Seligman and Teasdale, 1978) dealing with different ways in which people explain events in their lives. This approach builds upon locus of control, but extends it by stating that we also need to consider whether people attribute to stable causes or variable causes, and to global causes or specific causes.

Various scales have been developed to assess both attributional style and locus of control. Locus of control scales include those used by Rotter and later by Duttweiler, the Nowicki and Strickland (1973) Locus of Control Scale for Children and various locus of control scales specifically in the health domain, most famously that of Kenneth Wallston and his colleagues, The Multidimensional Health Locus of Control Scale (Wallston et al, 1978). Attributional style has been assessed by the Attributional Style Questionnaire (Peterson et al., 1982), the Expanded Attributional Style Questionnaire (Peterson & Villanova, 1988), the Attributions Questionnaire (Gong-guy & Hammen, 1990), the Real Events Attributional Style Questionnaire (Norman & Antaki, 1988) and the Attributional Style Assessment Test (Anderson, 1988).

Walter Mischel (1999) has also defended a cognitive approach to personality. His work refers to "Cognitive Affective Units", and considers factors such as encoding of stimuli, affect, goal-setting and self-regulatory beliefs. The term "Cognitive Affective Units" shows how his approach considers affect as well as cognition.

Albert Ellis, an American cognitive-behavioral therapist, is considered by many to be the grandfather of cognitive-behavioral therapy. In 1955 Ellis developed Rational Emotive Behavior Therapy (REBT), which later came to be known as Rational Therapy (RT). REBT required that the therapist help the client understand-and act on the understanding-that his personal philosophy contains common irrational beliefs that lead to his own emotional pain. Because thinking and emotion have a cause and effect relationship, Ellis believes that the thoughts we have become our emotions and the emotions we have become our thoughts. The basic theory of REBT is that majority of people create their own sort of emotional consequences because to sustain an emotion it must have had some form of thought. Ellis also created the A-B-C theory of personality. (A) is the activating event which is followed by (B), the belief system that the person holds and then (C), the emotional consequence. What the theory states is that (A) does not cause (C); but that (B) causes (C). The emotional consequences are caused by what the person believes in. An example would be if a person is walking outside and a stranger in a car pulls up next to them asking for directions (A), and the persons' belief system is that any stranger in a car that wants directions wants to hurt you (B) so therefore the person fears the person in the car is going to hurt them (C).

Aaron Beck, who is widely noted as the father of cognitive-behavioral therapy (CBT), suggested that nearly all psychological dilemmas can be redirected in a positive (helpful) manner with the changing of the suffering individual's thought processes. He has worked extensively on depression and suicide, and is now redirecting his theories towards those with borderline personality disorder, and the various anxiety disorders (OCD, neurosis, phobias, PTSD, etc.). Extensive evidence has proven the effectiveness of combining CBT with pharmacotherapy in treating the most severe psychiatric disorders such as bi-polar disorder and schizophrenia. Aaron Beck's continuing research in the field has proven to be a greater success over time.

Humanistic Theories: In humanistic psychology it is emphasized people have free will and that they play an active role in determining how they behave. Accordingly, humanistic psychology focuses on subjective experiences of persons as opposed to forced, definitive factors that determine behaviour. Abraham Maslow and Carl Rogers were proponents of this view.

Maslow spent much of his time studying what he called "self-actualizing persons", those who are "fulfilling themselves and doing the best that they are capable of doing". Maslow believes that all who are interested in growth move towards self-actualizing (growth, happiness, satisfaction) views. Many of these people demonstrate a trend in dimensions of their personalities. Characteristics of self-actualizers according to Maslow include the four key dimensions; (1) Awareness-maintaining constant enjoyment and awe of life.These individuals often experienced a "peak experience". He defined a peak experience as an "intensification of any experience to the degree that there is a loss or transcendence of self". A peak experience is one in which an individual perceives an expansion of his or herself, and detects a unity and meaningfulness in life. Intense concentration on an activity one is involved in, such as running a marathon, may invoke a peak experience. (2) Reality and problem centered-they have tendency to be concerned with "problems" in their surroundings. (3) Acceptance/ Spontaneity-they accept their surroundings and what cannot be changed. And (4) Unhostile sense of humor/democratic-they do not like joking about others, which can be viewed as offensive. They have friends of all backgrounds and religions and hold very close friendships.

Maslow and Rogers emphasized a view of the person as an active, creative, experiencing human being who lives in the present and subjectively responds to current perceptions, relationships, and encounters. They disagree with the dark, pessimistic outlook of those in the Freudian psychoanalysis ranks, but rather view humanistic theories as positive and optimistic proposals which stress the tendency of the human personality toward growth and self-actualization. This progressing self will remain the center of its constantly changing world; a world that will help mold the self but not necessarily confine it. Rather, the self has opportunity for maturation based on its encounters with this world. This understanding attempts to reduce the acceptance of hopeless redundancy. Humanistic therapy typically relies on the client for information of the past and its effect on the present, therefore the client dictates the type of guidance the therapist may initiate. This allows for an individualized approach to therapy. Rogers found that patients differ in how they respond to other people. Rogers tried to model a particular approach to therapy-he stressed the response. These responses came in a variety of fashions:

A. *Evaluative Response:* Place a value judgment on person's feelings.

B. *Interpretive Response:* Tells the person what they're really thinking or feeling.

C. *Reflective Response*: Captures how someone is feeling right now about the situation.

Biopsychological Theories: Around the 1990s, neuroscience entered the domain of personality psychology. Whereas previous efforts for identifying personality differences relied upon simple, direct, human observation, neuroscience introduced powerful brain analysis tools like Electroencephalography (EEG), Positron Emission Tomography (PET), and Functional Magnetic Resonance Imaging (fMRI) to this study. One of the founders of this area of brain research is Richard Davidson of the University of Wisconsin-Madison. Davidson's research lab has focussed on the role of the prefrontal cortex (PFC) and amygdala in manifesting human personality. In particular, this research has looked at hemispheric asymmetry of activity in these regions. Neuropsychological studies have illustrated how hemispheric asymmetry can affect an individual's personality (particularly in social settings) for individuals who have NLD (non-verbal learning disorder) which is marked by the impairment of nonverbal information controlled by the right hemisphere of the brain. Difficulties will arise in the areas of gross motor skills, inability to organize visual-spatial relations, or adapt to novel social situations. Frequently, a person with NLD is unable to interpret non-verbal cues, and therefore experiences difficulty interacting with peers in socially normative ways.

Personality Tests

There are two major types of personality tests. Projective tests assume that personality is primarily unconscious and assess an individual by how he or she responds to an ambiguous stimulus, like an ink blot. The idea is that unconscious needs will come out in the person's response, e.g. a very hostile person may see images of destruction. Objective tests assume that personality is consciously accessible and measure it by self-report questionnaires. Research on psychological assessment has generally found that objective tests are more valid and reliable than projective tests.

Examples of personality tests include:

- Holland Codes
- Rorschach test
- Minnesota Multiphasic Personality Inventory
- Morrisby Profile
- Myers-Briggs Type Indicator
- Enneagram Type Indicator

- NEO PI-R
- Thematic Apperception Test
- Kelly's Repertory Grid

Critics have pointed to the Forer effect to suggest that some of these appear to be more accurate and discriminating than they really are.

PHYSIOLOGICAL PSYCHOLOGY

Physiological psychology is a subdivision of biological psychology that studies the neural mechanisms of behavior through direct manipulation of the brains of nonhuman animal subjects in controlled experiments.

Unlike other subdivisions within biological psychology, the main focus of physiological psychological research is the development of theories that explain brain-behavior relationships rather than the development of research that has translational value.

One example of physiological psychology research is the study of the role of the hippocampus in learning and memory. This can be achieved by surgical removal of the hippocampus from the rat brain followed by an assessment of memory tasks by that same rat.

In the past, physiological psychologists received much of their their training in psychology departments in major universities. Currently, physiological psychologists are also be trained in behavioral neuroscience or biological psychology programs that are affiliated with psychology departments, or in interdisciplinary neuroscience programs.

SOCIAL PSYCHOLOGY

Social psychology is the scientific study of how people's thoughts, feelings, and behaviors are influenced by the actual, imagined, or implied presence of others (Allport, 1985). By this definition, scientific refers to the empirical method of investigation. The terms thoughts, feelings, and behaviors include all of the psychological variables that are measurable in a human being. The statement that others may be imagined or implied suggests that we are prone to social influence even when no other people are present, such as when watching television, or following internalized cultural norms.

Social psychologists typically explain human behavior as a result of the interaction of mental states and immediate, social situations. In Kurt Lewin's (1951), famous heuristic, behavior can be viewed as a function of the person and the environment, B=f(P,E). In general, social psychologists have a preference for laboratory based, empirical findings. Their theories tend to be specific and focused, rather than global and general.

Social psychology is an interdisciplinary domain that bridges the gap between psychology and sociology. During the years immediately following World War II, there was frequent collaboration between psychologists and sociologists (Sewell, 1989). However, the two disciplines have become increasingly specialized and isolated from each other in recent years, with sociologists focusing on "macro variables" (e.g. social structure) to a much greater extent. Nevertheless, sociological approaches to social psychology remain an important counterpart to psychological research in this area.

History

The discipline of social psychology began in the United States at the dawn of the 20th Century. The first published study in this area was an experiment by Norman Triplett (1898) on the phenomenon of social facilitation. During the 1930s, many Gestalt psychologists, particularly Kurt Lewin, fled to the United States from Nazi Germany. They were instrumental in developing the field as something separate from the behavioral and psychoanalytic schools that were dominant during that time, and social psychology has always maintained the legacy of their interests in perception and cognition. Attitudes and a variety of small group phenomena were the most commonly studied topics in this era.

During WWII, social psychologists studied persuasion and propaganda for the U.S. military. After the war, researchers became interested in a variety of social problems, including gender issues and racial prejudice. In the sixties, there was growing interest in a variety of new topics, such as cognitive dissonance, bystander intervention, and aggression. By the 1970s, however, social psychology in America had reached a crisis. There was heated debate over the ethics of laboratory experimentation, whether or not attitudes really predicted behavior, and how much science could be done in a cultural context. This was also the time when a radical situationist approach challenged the relevance of self and personality in psychology.

Social psychology reached maturity in both theory and method during the 1980s and 1990s. Careful ethical standards now regulate research, and greater pluralism and multicultural perspectives have emerged. Modern researchers are interested in a variety of phenomena, but attribution, social cognition, and the self-concept are perhaps the greatest areas of growth in recent years. Social psychologists have also maintained their applied interests, with contributions in health and environmental psychology, as well as the psychology of the legal system.

INTRAPERSONAL PHENOMENA

Attitudes: The study of attitudes is a core topic in social psychology. Attitudes are involved in virtually every other area of the discipline, including conformity, interpersonal attraction, social perception, and prejudice. In social psychology, attitudes are defined as learned, global evaluations of a person, object, place, or issue that influence thought and action (Perloff, 2003). Put more simply, attitudes are basic expressions of approval or disapproval, favorability or unfavorability, or as Bem (1970) put it, likes and dislikes. Examples would include liking chocolate ice cream, being anti-abortion, or endorsing the values of a particular political party.

Social psychologists have studied attitude formation, the structure of attitudes, attitude change, the function of attitudes, and the relationship between attitudes and behavior. Because people are influenced by the situation, general attitudes are not always good predictors of specific behavior. For a variety of reasons, a person may value the environment and not recycle a can on a particular day. Attitudes that are well remembered and central to our self-concept, however, are more likely to lead to behavior, and measures of general attitudes do predict patterns of behavior over time.

Persuasion: The topic of persuasion has received a great deal of attention in recent years. Persuasion is an active method of influence that attempts to guide people toward the adoption of an attitude, idea, or behavior by rational or emotive means. Persuasion relies on "appeals" rather than strong pressure or coercion. Numerous variables have been found to influence the persuasion process, and these are normally presented in four major categories: who said what to whom and how.

1. The Communicator, including credibility, expertise, trustworthiness, and attractiveness.
2. The Message, including varying degrees of reason, emotion (such as fear), one-sided or two sided arguments, and other types of informational content.
3. The Audience, including a variety of demographics, personality traits, and preferences.
4. The Channel, including thc printcd word, radio, television, the internet, or face-to-face interactions.

Dual process theories of persuasion (such as the Elaboration Likelihood Model) maintain that the persuasive process is mediated by two separate "routes." Persuasion can be accomplished by either superficial aspects of the communication or the internal logic of the message. Whether someone is persuaded by a popular celebrity or factual arguments is largely determined by the ability and motivation of the audience. However, decades of research have demonstrated that deeply held attitudes are remarkably resistant to persuasion under normal circumstances.

Social Cognition: Social cognition is a growing area of social psychology that studies how people perceive, think about, and remember information about others. One assumption in social cognition is that reality is too complex to easily discern, and so we see the world according to simplified schemas or images of reality. Schemas are generalized mental representations that organize knowledge and guide information processing. For example, one's schema for mice might include the expectation that they are small, and furry, and eat cheese.

Schemas often operate automatically and unintentionally, and can lead to biases in perception and memory. Schematic expectations may lead us to see something that is not there. One experiment found that people are more likely to misperceive a weapon in the hands of a black man than a white man (Correll, et al., 2002). This type of schema is actually a stereotype, a generalized set of beliefs about a particular group of people. Stereotypes are often related to negative or preferential attitudes (prejudice) and behavior (discrimination). Schemas for types of events (e.g. going to McDonalds, doing laundry) are known as scripts.

Another major concept in social cognition is attribution. Attributions are the explanations we make for people's behavior, either our own behavior or the behavior of others. An attribution can be either internal or external. Internal or dispositional attributions assign causality to factors within the person, such as ability or personality. External or situational attributions assign causality to an outside factor, such as the weather. Numerous biases in the attribution process have been discovered:

- Fundamental attribution error-the tendency to make dispositional attributions for behavior. The actor-observer effect is a refinement of this bias, the tendency to make dispositional attributions for other people's behavior and situational attributions for our own.
- Just world effect-the tendency to blame victims (a dispositional attribution) for their suffering. This is believed to be motivated by people's anxiety that good people, including themselves, could be victimized in an unjust world.
- Self-serving bias-the tendency to take credit for successes, and blame others for failure. Researchers have found that depressed individuals often lack this bias and actually have more realistic perceptions of reality.

Heuristics are cognitive short cuts. Instead of weighing all the evidence when making a decision, people rely on heuristics to save time and energy. The availability heuristic occurs when people estimate the probability of an outcome based on how easy that outcome is to imagine. As such, vivid or highly memorable possibilities will be perceived as more likely than those that are-harder to picture or are difficult to understand, resulting in a corresponding cognitive bias.

There are a number of other biases that have been found by social cognition researchers. The hindsight bias is a false memory of having predicted events, or an exaggeration of actual predictions, after becoming aware of the outcome. The confirmation bias is a type of bias leading to the tendency to search for, or interpret information in a way that confirms one's preconceptions.

Self-concept: The fields of social psychology and personality have merged over the years, and social psychologists have developed an interest in a variety of self-related phenomena. In contrast with traditional personality theory, however, social psychologists place a greater emphasis on cognitions than on traits. Much research focuses on the self-concept, which is a person's understanding of his or her self. The self-concept can be divided into a cognitive component, known as the self-schema, and an evaluative component, the self-esteem. The need to maintain a healthy self-esteem is recognized as a central human motivation in the field of social psychology. Self-efficacy beliefs are an aspect of the self-schema. Self-efficacy refers to an individual's expectation that performance on some task will be effective and successful.

People develop their self-concepts by a variety of means, including introspection, feedback from others, self-perception, and social comparison. By comparison to relevant others, people gain information about themselves, and they make inferences that are relevant to self-esteem. Social comparisons can be either upward or downward, that is, comparisons to people who are either higher in status or ability, or lower in status or ability. Downward comparisons are often made in order to elevate self-esteem.

Self-perception is a specialized form of attribution that involves making inferences about oneself after observing one's own behavior. Psychologists have found that too many extrinsic rewards (e.g. money) tend to reduce intrinsic motivation through the self-perception process. People's attention is directed to the reward and they lose interest in the task when the reward is no longer offered. This is an important exception to reinforcement theory.

Cognitive Dissonance: Cognitive dissonance is a feeling of unpleasant arousal caused by noticing an inconsistency among one's cognitions (Festinger, 1957). Cognitive dissonance was originally developed as a theory of attitude change, but it is now considered to be a self theory by most social psychologists. Dissonance is strongest when a discrepancy has been noticed between one's self-concept and one's behavior, e.g. doing something that makes one ashamed. This can result in self-justification as the individual attempts to deal with the threat. Cognitive dissonance typically leads to a change in attitude, a change in behavior, a self-affirmation, or a rationalization of the behavior.

An example of cognitive dissonance is smoking. Smoking cigarettes increases the risk of cancer, which is threatening to the self-concept of the individual who smokes. Most of us believe ourselves to be intelligent and rational, and the idea of doing something foolish and self-destructive causes dissonance. To reduce this uncomfortable tension, smokers tend to make excuses for themselves, such as "I'm going to die anyway, so it doesn't matter."

INTERPERSONAL PHENOMENA

Social Influence: Social influence refers to the way people affect the thoughts, feelings, and behaviors of others. Like the study of attitudes, it is a traditional, core topic in social psychology. In fact, research on social influence overlaps considerably with research on attitudes and persuasion. Social influence is also closely related to the study of group dynamics, as most of the principles of influence are strongest when they take place in social groups.

Conformity is the most common and pervasive form of social influence. It is generally defined as the tendency to act or think like other members of a group. Group size, unanimity, cohesion, status, and prior commitment all help to determine the level of conformity in an individual. Conformity is usually viewed as a negative tendency in American culture, but a certain amount of conformity is not only necessary and normal, but probably essential for a community to function.

The two major motives in conformity are: (1) normative influence, the tendency to conform in order to gain social acceptance, and avoid social rejection or conflict, as in peer pressure; and (2) informational influence, which is based on the desire to obtain useful information through conformity, and thereby achieve a correct or appropriate result. Minority influence is the degree to which a smaller faction within the group influences the group during decision making. Note that this refers to a minority position on some issue, not an ethnic minority. Their influence is primarily informational and depends on consistent adherence to a position, degree of defection from the majority, and the status and self-confidence of the minority members. Reactance is a tendency to assert oneself by doing the opposite of what is expected. This phenomenon is also known as anticonformity and it appears to be more common in men than in women.

There are two other major areas of social influence research. Compliance refers to any change in behavior that is due to a request or suggestion from another person. The Foot-in-the-door technique is a compliance method in which the persuader requests a small favor and then follows up with a larger favor, e.g. asking for the time, and then asking for ten dollars. A related trick is the Bait and switch (Cialdini, 2000). The third major form of social influence is obedience. This is a change in behavior that is the result of a direct order or command from another person.

A different kind of social influence is the self-fulfilling prophecy. This is a prediction that, in being made, actually causes itself to become true. For example, in the stock market, if it is widely believed that a crash is imminent, investors may lose confidence, sell most of their stock, and actually cause the crash. Likewise, people may expect hostility in others and actually induce this hosility by their own behavior.

Group Dynamics: A group is two or more people that interact, influence each other, and share a common identity.

Groups have a number of emergent qualities that distinguish them from aggregates:

- Norms-implicit rules and expectations for group members to follow, e.g. saying thank you, shaking hands.
- Roles-implicit rules and expectations for specific members within the group, e.g. the oldest sibling, who may have additional responsibilities in the family.
- Relations-patterns of liking within the group, and also differences in prestige or status, e.g. leaders, popular people.

Temporary groups and aggregates share few or none of these features, and do not qualify as true social groups. People waiting in line to get on a bus, for example, do not constitute a group.

Groups are important not only because they offer social support, resources, and a feeling of belonging, but because they supplement an individual's self-concept. To a large extent, we define ourselves by our group memberships. This natural tendency for people to identify themselves with a particular group and contrast themselves with other groups is known as social identity (Tajfel & Turner, 1986). Unfortunately, social identity can lead to feelings of "us and them." It is frequently associated with preferential treatment toward the ingroup, and prejudice and discrimination against outgroups.

Groups often moderate and improve decision making, and are frequently relied upon for these benefits, such as committees and juries. A number of group biases, however, can interfere with effective decision making. For example, group polarization, formerly known as the risky shift, occurs when people polarize their views in a more extreme direction after group discussion. Even worse is the phenomenon of groupthink. This is a collective thinking defect that is characterized by a premature consensus. Groupthink is caused by a variety of factors, including isolation and a highly directive leader. Janis (1972) offered the 1961 Bay of Pigs invasion as a historical case of groupthink.

Groups also affect performance and productivity. Social facilitation, for example, is a tendency to work harder and faster in the presence of others. Social facilitation increases the likelihood of the dominant response, which tends to improve performance on simple tasks and reduce it on complex tasks. In contrast, social loafing is the tendency of individuals to slack when working in a group. Social loafing is common when the task is considered unimportant and individual contributions are not easy to see.

Social psychologists study a variety of group related, or collective phenomena such as the behavior of crowds. An important concept in this area is deindividuation, a reduced state of self-awareness that can be caused by feelings of anonymity. Deindividuation is associated with uninhibited and sometimes dangerous behavior. It is common in crowds and mobs, but it can also be caused by a disguise, a uniform, alcohol, dark environments, or online anonymity.

Relations with Others: Social psychologists are interested in the question of why people sometimes act in a prosocial way (helping, liking, or loving others), but at other times act in an antisocial way (hostility, aggression, or prejudice against others).

Aggression can be defined as any behavior that is intended to harm another human being. Hostile aggression is accompanied by strong emotions, particularly anger. Harming the other person is the goal. Instrumental aggression is only a means to an end. Harming the person is used to obtain some other goal, such as money. Research indicates that there are many causes of aggression, including biological factors like testosterone and environmental factors, such as social learning. Immediate situational factors such as frustration are also important in triggering an aggressive response.

Although violence is a fact of life, people are also capable of helping each other, even complete strangers in emergencies. Research indicates that altruism occurs when a person feels empathy for another individual, even in the absence of other motives (Batson, 1998). However, according to the bystander effect, the probability of receiving help in an emergency situation drops as the number of bystanders increases. This is due to conformity effects and a diffusion of responsibility (Latane, 1981).

Interpersonal Attraction: Another major area in the study of people's relations to each other is interpersonal attraction. This refers to all of the forces that lead people to like each other, establish relationships, and in some cases, fall in love. Several general principles have been discovered by researchers in this area:

- Proximity-physical proximity increases attraction, as opposed to long distance relationships which are more at risk.
- Familiarity-mere exposure to others increases attraction, even when the exposure is not consciously realized.
- Similarity-the more similar two people are in attitudes, background, and other traits, the more probable it is that they will like each other. Contrary to popular opinion, opposites do not usually attract.

Physical attractiveness is an important element of romantic relationships, particularly in the early stages which are characterized by high levels of passion. Later on, similarity becomes more important and the type of love people experience shifts from passionate to companionate. Robert Sternberg (1986) has suggested that there are actually three components to love: intimacy, passion, and commitment.

According to social exchange theory, relationships are based on rational choice and cost-benefit analysis. If one partner's costs begin to outweigh his or her benefits, that person may leave the relationship, especially if there are good alternatives available. With time, long term relationships tend to become communal rather than simply based on exchange.

POSITIVE PSYCHOLOGY

Positive psychology is the scientific study of optimal human functioning. People have been discussing the question of human happiness since at least Ancient Greece.

The modern history of psychology has been criticized as primarily dedicated to addressing mental illness rather than mental "wellness", as early as those of the American psychologist and philosopher William James. In his 1902 book, The Varieties of Religious Experience, James argues that happiness is a chief concern of human life and those who pursue it should be regarded as "healthy-minded." Several humanistic psychologists-such as Abraham Maslow, Carl Rogers, and Erich Fromm-developed successful theories and practices that involved human happiness despite there being a lack of solid empirical evidence behind their work.

Modern empirical researchers in this subfield include Don Clifton, Albert Bandura, Martin Seligman, Ed Diener, Mihaly Csikszentmihalyi, C.R. Snyder, Christopher Peterson, Shelley Taylor, Barbara Fredrickson.

Research: Some researchers in this field posit that positive psychology can be delineated into three overlapping areas of research:

1. Research into the Pleasant Life or the "life of enjoyment" examines how people optimally experience, forecast, and savor the positive feelings and emotions that are part of normal and healthy living (e.g. relationships, hobbies, interests, entertainment, etc.).
2. The study of the Good Life or the "life of engagement" investigates the beneficial affects of immersion, absorption, and flow that individuals feel when optimally engaged with their primary activities. These states are experienced when there is a positive match between a person's strength and the task they are doing, i.e. when they feel confident that they can accomplish the tasks they face.
3. Inquiry into the Meaningful Life or "life of affiliation" questions how individuals derive a positive sense of well-being, belonging, meaning, and purpose from being part of and contributing back to something larger and more permanent than themselves (e.g. nature, social groups, organizations, movements, traditions, belief systems).

These categories appear to be neither widely disputed nor adopted by researchers across this relatively new academic field.

Application: The development of the Character Strengths and Virtues (CSV) handbook represents the first attempt on the part of the research community to identify and classify the positive psychological traits of human beings. Much like the Diagnostic and Statistical Manual of Mental Disorders (DSM) of general psychology, the CSV provides a theoretical framework to assist in developing practical applications for positive psychology. This manual identifies six classes of virtue (i.e. "core virtues"), made up of twenty-four measurable character strengths. The organization of these virtues and strengths is as follows:

1. Wisdom and Knowledge: creativity, curiosity, open-mindedness, love of learning, perspective
2. Courage: bravery, persistence, integrity, vitality
3. Humanity: love, kindness, social intelligence
4. Justice: citizenship, fairness, leadership
5. Temperance: forgiveness and mercy, humility and modesty, prudence, self-regulation
6. Transcendence: appreciation of beauty and excellence, gratitude, hope, humor, spirituality

Practical applications of positive psychology include helping individuals and organizations correctly identify their strengths and use them to increase and sustain their respective levels of well-being. Therapists, counselors, coaches, and various other psychological professionals can use the new methods and techniques to build and broaden the lives of individuals who are not necessarily suffering from mental illness or disorder.

PSYCHOPATHOLOGY

Psychopathology is a term which refers to either

1. the study of mental illness or mental distress
2. the manifestation of behaviours and experiences which may be indicative of mental illness or psychological impairment.

It is also the name of an academic journal that specialises in the understanding and classification of mental illness in clinical psychiatry.

Psychopathology as the Study of Mental Illness: Many different professions may be involved in studying mental illness or distress. Most notably, psychiatrists and clinical psychologists are particularly interested in this area and may either be involved in clinical treatment of mental illness, or research into the origin, development and manifestations of such states, or often, both. More widely, many different specialties may be involved in the study of psychopathology. For example, a neuroscientist may focus on brain changes related to mental illness. Therefore, someone who is referred to as a psychopathologist, may be one of any number of professions who have specialised in studying this area.

Psychiatrists in particular are interested in descriptive psychopathology, which has the aim of describing the symptoms and syndromes of mental illness. This is both for the diagnosis of individual patients, or for the creation of diagnostic systems (such as the Diagnostic and Statistical Manual of Mental Disorders) which define exactly which signs and symptoms should make up a diagnosis, and how experiences and behaviours should be grouped in particular diagnoses (e.g. clinical depression, schizophrenia).

Psychopathology as a Descriptive Term: The term psychopathology may also be used to denote behaviours or experiences which are indicative of mental illness, even if they do not constitute a formal diagnosis. For example, the presence of an hallucination may be considered as a psychopathological sign, even if there are not enough symptoms present to fulfill the criteria for one of the disorders listed in the DSM. In a more general sense, any behaviour or experience which causes impairment, distress or disability, particularly if it is thought to arise from a functional breakdown in either the cognitive and neurocognitive systems in the brain, may be classified as psychopathology.

The Academic Journal Psychopathology: Originally founded in 1897 and named Psychiatria Clinica, the journal changed its name to Psychopathology in 1984. It bills itself as the 'International journal of experimental psychopathology, phenomenology and psychiatric diagnosis' and aims to 'elucidate the complex interrelationships of biology, subjective experience, behavior and therapies'.

4

Psychological Measures

Psychological testing or psychological assessment is a field characterized by the use of samples of behavior in order to infer generalizations about a given individual. The technical term for the science behind psychological testing is psychometrics. By samples of behavior, one means observations over time of an individual performing tasks that have usually been prescribed beforehand. These responses are often compiled into statistical tables that allow the evaluator to compare the behavior of the individual being tested to the responses of a norm group. When multiple tests are administered, the procedure is referred to as full battery assessment.

Psychological testing is not the same as psychological assessment. Psychological assessment uses tests and other information such as personal and medical history, description of current symptoms and problems by either self or others, and collateral information (interviews with other persons about the person being assessed). A useful psychological measure must be both valid (i.e., actually measures what it claims to measure) and reliable (i.e., internally consistent or give consistent results over time).

TYPES OF PSYCHOLOGICAL EVALUATIONS

There are several broad categories of psychological evaluation tests:

Norm-referenced

Norm-referencing tests regard a specific or non-specific topic, often pertaining to the test taker's education, or career.

These tests are commonly used in high schools and colleges. The results often correlate to a Bell curve--with a few participants doing badly, most doing average, and a few doing extremely well. A common norm-referencing test in the United States is the Graduate Record Examination (GRE), which measures the general area of critical thinking, verbal reasoning and quantitative reasoning skills. In addition, specific GRE subject tests however cover eight broad fields, mainly in the sciences, but also in mathematics and English.

IQ Tests

IQ (or cognitive) tests and academic achievement tests are the most common norm-referenced tests. In either of these types of tests, a series of tasks is presented to the person being evaluated, and the person's responses are graded according to carefully prescribed guidelines. After the test is completed, the results can be compiled and compared to the responses of a norm group, usually comprised of people at the same age or grade level as the person being evaluated.

IQ tests (e.g., WAIS-III, WISC-IV, Cattell Culture Fair III, K-BIT (Kaufman & Kaufman, 1990), Universal Nonverbal Intelligence Test) and academic achievement tests (e.g., WIAT, WRAT) are designed to be administered to either an individual (by a trained evaluator) or to a group of people (paper and pencil tests). The individually-administered tests tend to be more comprehensive, more reliable, more valid and generally to have better psychometric characteristics than group-administered tests. However, individually-administered tests are more expensive to administer because of the need for a trained administrator (psychologist, school psychologist, or psychometrician) and because of the limitation of working with just one client at a time.

Neuropsychological Tests

These tests consist of specifically designed tasks used to measure a psychological function known to be linked to a particular brain structure or pathway. They are typically used to assess impairment after an injury or illness known to affect neurocognitive functioning, or when used in research, to contast neuropsychological abilities across experimental groups.

Personality Tests

Psychological measures of personality are often described as either objective tests or projective tests.

Objective Tests (Rating Scale)

Objective tests have a restricted response format, such as allowing for true or false answers or rating using an ordinal scale. Prominent examples of objective personality tests include the Minnesota Multiphasic Personality Inventory, Millon Clinical Multiaxial Inventory-III (Millon, 1994), Child Behavior Checklist (Achenbach & Rescorla, 2001), and the Beck Depression Inventory (Beck & Steer, 1996). Objective personality tests can be designed for use in business for potential employees, such as the NEO-PI, the 16PF, and the Occupational Personality questionnaire, all of which are based on the Big Five taxonomy. The Big Five, or Five Factor Model of normal personality has gained acceptance since the early 1990s when some influential meta-analyses (e.g., Barrick & Mount 1991) found consistent relationships between the Big Five factors.

Projective Tests (Free Response Measures)

Projective tests allow for a much freer type of response. An example of this would be the Rorschach test, in which a person states what each of ten ink blots might be. The terms "objective test" and "projective test" have recently come under criticism in the Journal of Personality Assessment.

The more descriptive "rating scale or self-report measures" and "free response measures" are suggested, rather than the terms "objective tests" and "projective tests," respectively. There remains some controversy regarding the utility and validity of projective testing which is based on Freud's concept of projecting one's own personality attributes onto a neutral stimulus. However, many practitioners continue to rely on projective testing, and some testing experts (e.g., Cohen, Anastasi) suggest that these measures can be useful in developing therapeutic rapport.

The most widely used scoring system for the Rorschach is Exner's Comprehensive System (Exner,). Another common projective test is the Thematic Apperception Test (Murray, 1943, which is often scored with Drew Westen's (1991) Social Cognition and Object Relations Scales and Phebe Cramer's Defense Mechanisms Manual (1991, 2002). Both "rating scale" and "free response" measures are used in contemporary clinical practice. Other projective tests include the House-Tree-Person Projective Personality Test and the Thematic Apperception Test (TAT).

Direct Observation Tests

Although most psychological tests are "rating scale" or "free response" measures, psychological assessment may also involve the observation of people as they complete activities. This type of assessment is usually conducted with families in a laboratory, home or with children in a classroom. The purpose may be clinical, such as to establish a pre-intervention baseline of a child's hyperactive or aggressive classroom behaviors or to observe the nature of a parent-child interaction in order to understand a relational disorder. Direct observation procedures are also used in research, for example to study the relationship between intrapsychic variables and specific target behaviors, or to explore sequences of behavioral interaction.

The Parent-Child Interaction Assessment-II(PCIA; Holigrocki, Kaminski & Frieswyk, 1999) is an example of a direct observation procedure that is used with school-age children and parents. The parents and children are videotaped playing at a make-believe zoo. The Parent-Child Early Relational Assessment (Clark, 1999) is used to study parents and young children and involves a feeding and a puzzle task. The MacArthur Story Stem Battery(MSSB; Bretherton et al., 1990) is used to elicit narratives from children. The Dyadic Parent-Child Interaction Coding System-II(Eyberg, 1981) tracks the extent to which children follow the commands of parents and vice versa and is well suited to the study of children with Oppositional Defiant Disorders and their parents.

Psychological Evaluations using Data Mining

An examiner may use data mining methods to draw inferences from existing records, texts, and datasets about the person. One such technique is the Abika Test. This test involves gathering data on the individual such as public records, behavior history records, consumer activities, shopping histories, memberships in various organizations, court records, demographic data, property deeds, media, public and private databases, newsgroups, opinions expressed in chat rooms, forums, message boards including other methods such as statistical comparisons with peer groups, polling and information submitted by friends, co-workers, relatives. Thus, this technique compiles personality and psychological profiles that are based on inferences developed from existing records.

CLINICAL PSYCHOLOGY

Clinical psychology is the application of psychology to prevent and relieve psychologically-based distress or dysfunction and to promote subjective well-being. Clinical psychologists assess mental health problems, conduct and use scientific research to understand mental health problems, and develop, provide and evaluate psychological care and interventions (psychotherapy). Although different countries require various educational qualifications to practice clinical psychology, it has traditionally required a Doctorate degree-such as the PhD or PsyD in America or the DClinPsy in Britain-although the U.S. now has accredited Masters-level programs as well. Clinical psychology's therapeutic practices can address issues with individuals, couples, children, families, small groups, and communities.

The term was introduced in a 1907 paper by the American psychologist Lightner Witmer, who specifically defined it as the study of individuals, by observation or experimentation, with the intention of promoting change. The American Psychological Association offers a more modern definition of clinical psychology: The field of Clinical Psychology integrates science, theory, and practice to understand, predict, and alleviate maladjustment, disability, and discomfort as well as to promote human adaptation, adjustment, and personal development. Clinical Psychology focuses on the intellectual, emotional, biological, psychological, social, and behavioral aspects of human functioning across the life span, in varying cultures, and at all socioeconomic levels.

History

Clinical psychology developed partly as a result of a need for additional clinicians to treat mental health problems, and partly as psychological science advanced to the stage where the fruits of psychological research could be successfully applied in clinical settings. The field as it exists today can be said to have begun with Witmer's establishment of the first psychological clinic in 1896 at the University of Pennsylvania. He also founded the first journal of clinical psychology, Psychological Clinic.

Witmer's call for clinical involvement by psychologists was slow to gain acceptance, but by 1914 there were twenty-six more clinics in the United States. While Whitmer focused on individuals with intellectual deficits, others focused on those in mental distress, and clinical psychology was developing in mental hospitals as psychologists gained staff positions, often working alongside psychiatrists.

In the early 20th century, the work of Sigmund Freud and Josef Breuer-whilst not explicitly clinical psychology-gave great impetus to psychological understandings of mental distress and disorder. The assessment-based focus of early clinical psychology came fully into its own during World War I when the U.S. military required clinical psychologists to assess thousands of new soldiers.

It wasn't until 1917 that clinical psychologists began to organize under that name, when the American Association of Clinical Psychology was organized, leading to the Section on Clinical Psychology of the American Psychological Association. This section certified clinical psychologists until 1927, but would not allow clinical psychologists full membership in APA.

Slow growth of the field continued in the 1930s as several scattered, applied psychological organizations in the U.S. formed an alliance under the American Association of Applied Psychology. This group became the primary forum for clinical and applied psychology in the U.S. until APA reorganized in 1945, creating its division of clinical psychology, today known as Division 12.

Before the 1940s, individual psychotherapy was mostly conducted by psychiatrists, leaving clinical psychologists to focus on assessment. This changed during World War II, however, when the military gave greater recognition to the condition they termed "shell shock" which eventually came to be called Post Traumatic Stress Disorder (PTSD). The need for treatment was such that military called upon clinical psychologists to assist.

After WWII, a similar problem was faced when tens of thousands of soldiers came home needing psychological care. To meet this challenge in the U.S., the Veterans Administration made an enormous investment to set up programs to train doctoral-level clinical psychologists. As a consequence, the U.S. went from having no formal university programs in clinical psychology in 1946 to over half of all PhDs in psychology in 1950 being awarded in clinical psychology. In 1947, a report was drafted that led to the scientist/practitioner model of clinical psychology, known today as the Boulder Model. This called on clinical psychologists to train as scientific psychologists as well as focusing on interpersonal clinical skills. Similar organizational and theoretical developments took place in other countries in the 1950s. The number of clinical psychologists and academic journals proliferated.

Different waves of theory influenced the field after WWII. First Freudian and neo-freudian (or psychodynamic), then humanistic, and finally orientations founded in behaviorism and cognitivism, eventually leading to Cognitive Behavioral Therapy. More recently, orientations founded in transpersonal psychology are becoming more common. Clinical psychology has remained inclusive of an eclectic range of theoretical perspectives and practical methods.

Training

The University of Pennsylvania was the first to offer formal education in clinical psychology.

Doctoral Level Training

Clinical psychologists typically undergo many hours of postgraduate training-usually 4 to 6 years post-Bachelors-under supervision in order to gain demonstrable competence and experience. Today, in America, about half of the licensed psychologists are being trained in the Scientist-Practitioner Model of Clinical Psychology (PhD)-a model that emphasizes research and is usually housed in universities. The other half are being trained within a Practitioner-Scholar Model of Clinical Psychology (PsyD), which has more focus on practice (similar to professional degrees for medicine and law).

Both models envision practicing Clinical Psychology in a research-based, scientifically valid manner. The American Psychological Association, among many English-speaking psychological societies, supports both models and encourages accreditation of PhD and PsyD programs that meet its strict academic standards.

Doctorate (PhD and PsyD) programs usually involve some variation on the following 4 to 6 year, 90-unit curriculum:

- Bases of behavior-biological, cognitive-affective, and cultural-social
- Individual differences-personality, lifespan development, psychopathology
- History and systems-development of psychological theories, practices, and scientific knowledge
- Clinical practice-diagnostics, psychological assessment, psychotherapeutic interventions, psychopharmacology, ethical and legal issues
- Clinical experience
    - Practicum-usually one or two years of working with clients under supervision in a clinical setting
    - Doctoral Internship-usually an intensive one or two year placement in a clinical setting
- Dissertation-PhD programs usually require original quantitative empirical research, while PsyD dissertations often address qualitative research, theoretical scholarship, program evaluation or development, critical literature analysis, or clinical application and analysis
- Specialized electives-many programs offer sets of elective courses for specializations, such as health, child, family, community, or neuropsychology

- Personal psychotherapy-many programs require students to undertake a certain number of hours of personal psychotherapy (with a non-faculty therapist).

Masters Level Training

There are a number of U.S. schcols offering accredited programs in clinical psychology resulting in a Masters degree. Such programs can range from 48 to 84 units, most often taking 2 to 3 years to complete post-Bachelors. Training usually emphasizes theory and treatment over research, quite often with a focus on school or couples and family counseling. Similar to doctoral programs, Masters-level students usually must fulfill time in a clinical practicum under supervision and undergo a minimum amount of personal psychotherapy. While many graduates from Masters-level training go on to doctoral programs, a large number also go directly into practice-often as a Licensed Professional Counselor (LPC), Marriage and Family Therapist (MFT) or other similar license.

Training in Britain

In Britain, clinical psychologists undertake a DClinPsy (or similar) which is a taught doctorate with both clinical and research components. This is a three-year full-time course, taken after a three-year undergraduate degree and some form of experience, usually in either the National Health Service as an Assistant Psychologist or in academia as a Research Assistant. Previously the training required obtaining an MSc and/or MPhil degree, with some also undertaking a PhD and some going on to academic study.

Professional Practice

Clinical psychologists can offer a range of professional services, including:

- Provide psychological treatment (psychotherapy)
- Administer and interpret psychological assessment and testing
- Conduct psychological research
- Teach

- Development of prevention programs
- Consultation (especially with schools and businesses)
- Program administration
- Provide expert testimony (forensics).

In practice, clinical psychologists may work with individuals, couples, families, or groups in a variety of settings, including private practices, hospitals, mental health organizations, schools, businesses, and non-profit agencies. Most (but not all) clinical psychologists who engage in research and teaching do so within a college or university setting. Clinical psychologists may also choose to specialize in a particular field-common areas of specialization, some of which can earn board certification, include:

- Specific disorders (trauma, addiction, eating, sleep, sex, depression, anxiety, phobias, etc.)
- Neuropsychological disorders
- Child and adolescent
- Family and relationship counseling
- Health
- Sport
- Forensic
- Organization and business
- School.

Clinical Psychology and other Mental Health Professions

Psychiatry

Although clinical psychologists and psychiatrists share the same fundamental aim-the alleviation of mental distress-their training, outlook, and methodologies are often quite different. Perhaps the most significant difference is that psychiatrists are medical doctors with 4 years of medical school and another 4 years of residency in a medical setting where they can often choose to specialize, such as working with children or people with specific conditions. As such, they tend to use the medical model to assess psychological problems (i.e. those they treat are seen as patients with an illness) and rely on psychotropic medications as the chief method of addressing them (although many also employ psychotherapy as well). Their medical training does give them an advantage in terms of being able to conduct physical examinations, order and interpret laboratory tests and EEGs, and may order brain imaging studies such as CT or CAT, MRI, and PET scanning.

Clinical psychologists do not usually prescribe medication, although there is a growing movement for psychologists to have limited prescribing privileges. Such privileges require additional, supervised training and education, and would mostly be limited to psychotropic medications. To date, psychologists who obtain additional training in clinical psychopharmacology may prescribe psychotropic medications in Guam, New Mexico, and Louisiana. In general, however, when medication is warranted many psychologists will work in cooperation with psychiatrists so that clients get all their therapeutic needs met.

Unless a psychiatrist voluntarily chooses to get extra training, such as at a psychoanalytic institute, they will have less training in the theory and practice of psychotherapy than will a licensed clinical psychologist. Even though many psychiatrists do seek out such training, the majority of them increasingly focus on medication management, possibly because insurance tends to pay far more for this service than for psychotherapy. Further, psychologists tend to have far more training in psychological assessment.

Counseling Psychology

Counseling generally involves helping people with what might be considered "normal" or "moderate" psychological problems, such as the feelings of anxiety or sadness resulting from major life changes or events. As such, counseling psychologists often help people adjust to or cope with their environment or major events, although many also work with more serious problems as well. Clinical psychologists, by comparison, are trained to help with these kinds of issues but also more debilitating or chronic problems, such as forms of dementia or psychosis. Other differences include: there are fewer counseling psychology graduate programs, they are usually housed in departments of education (as opposed to psychology departments for clin-psy programs), counseling psychologists tend to conduct more vocational assessment and less projective or objective assessment, and they are more likely to work in public service or university clinics (compared with clinical psychologists who are more likely to work in hospitals or private practice). Despite these differences, there is considerable overlap between the two fields and distinctions between them continue to fade.

School Psychology

School psychologists' primary concern is with the academic, social, and emotional well-being of children within a scholastic environment. Unlike clinical psychologists, they receive much more training in education, child development and behavior, and the psychology of learning, often graduating with a post-Masters Educational Specialist Degree (EdS) or Doctor of Education (EdD) degree. Besides offering individual and group therapy with children and their families, school psychologists also evaluate school programs, provide cognitive assessment, help design prevention programs (e.g. reducing drops outs), and work with teachers and administrators to help maximize teaching efficacy, both in the classroom and systemically.

Licensure

The practice of clinical psychology requires a license in the United States, Canada, the United Kingdom, and many other countries. Although each of the U.S. states is somewhat different in terms of requirements and licenses, there are three common elements:

Most states also require a certain number of continuing education credits per year in order to renew a license, which can be obtained though various means, such as taking audited classes and attending approved workshops.

Exact requirements vary by license and by state, and are commonly quite complex.

All U.S. state and Canada province licensing boards are members of the Association of State and Provincial Psychology Boards (ASPPB) which created and maintains the Examination for Professional Practice in Psychology (EPPP). Many states require other examinations in addition to the EPPP, such as a jurisprudence (i.e. mental health law) examination and/or an oral examination.

There are several licenses that allow one to practice clinical psychology in various forms, usually awarded in relation to one's educational degree.

- *Psychologist:* To practice with the title of Psychologist, in almost all cases a Doctorate degree is required (a PhD or PsyD in the U.S.). In addition to finishing 3 to 5 years of coursework, other requirements generally include finishing a dissertation, completing a pre-doc internship, fulfilling a certain number of supervised postdoctoral hours (usually taking 1 to 2 years), and passing the EPPP and any other provincial exams.
- *Marriage and Family Therapist (MFT):* An MFT license requires a Doctorate or Masters degree (which usually takes 2 to 3 years after a Bachelors and includes an internship and sometimes completion of a thesis). In addition, it usually involves 2 years of post-degree clinical experience under supervision, and licensure requires passing a written exam, commonly the National Examination for Marriage and Family Therapists which is maintained by the American Association for Marriage and Family Therapy. In addition, most state require an oral exam. MFTs, as the title implies, work mostly with families and couples, addressing a wide range of common psychological problems.
- *Licensed Professional Counselor (LPC)*: Similar to the MFT, the LPC license requires a Masters or Doctorate degree, a minimum number of hours of supervised clinical experience in a pre-doc practicum, and the passing of the National Counselor Exam. Similar licenses are the Licensed Mental Health Counselor (LMHC), Licensed Clinical Professional Counselor (LCPC), and Clinical Counselor in Mental Health (CCMH). In some states, after passing the exam, a temporary LPC license is awarded and the clinician may begin the normal 3000-hour supervised internship leading to the full

license allowing for the practice as a counselor or psychotherapist, usually under the supervision of a licensed psychologist.

- *Licensed Psychological Associate (LPA):* About twenty-six states offer a Masters-only license, a common one being the LPA, which allows for the therapist to either practice independently or (more commonly) under the supervision of a licensed psychologist, depending on the state. Common requirements are 2 to 4 years of post-Masters supervised clinical experience and passing a Psychological Associates Examination. Other titles for this level of licensing include Psychological Technician (Alabama), Psychological Assistant (California), Licensed Clinical Psychotherapist (Kansas), Licensed Psychological Practitioner (Minnesota), Licensed Behavioral Practitioner (Oklahoma), or Psychological Examiner (Tennessee).

Assessment: An important area of expertise for many clinical psychologists is psychological assessment, and there are indications that as much as 91% of psychologists engage in this core clinical practice. Such evaluation is usually done in service to gaining insight into and forming hypotheses about psychological or behavioral problems. As such, the results of such assessments are usually used to create generalized impressions rather than diagnoses.

There exists literally hundreds of various assessment tools, although only a few have been shown to have both high validity (i.e., test actually measures what it claims to measure) and reliability (i.e., test is consistent-internally, over time, and regardless of administrator). These measures generally fall within one of several categories, including the following:

- *Intelligence and Achievement Tests:* These tests are designed to measure certain specific kinds of cognitive functioning (often refered to as IQ) in comparison to a norming-group. Commonly used today are the Weschler tests (the WAIS-III for adults, the WISC-IV for children, and the WIAT-II achievement test), the Woodcock-Johnson-III, and the Stanford-Binet-5. These tests generally measure areas such as verbal skills (e.g. comprehention and vocabulary), memory (short and long term), attention span, arithmatic, and non-verbal performance (e.g. visual/spacial perception, hand-eye coordination, problem solving, and logical reasoning). These tests have been shown to accurately predict certain kinds of performance, especially scholastic.

- *Personality Tests:* Tests of personality aim to describe characteristic patterns of behavior, thoughts, and feelings that remain relatively stable throughout a person's lifetime. They generally fall within two categories: objective (offering restricted, measured responses, such as yes/no, true/false, or a rating scale) and projective (which allow a person to respond to ambiguous stimuli, presumably revealing non-conscious psychological dynamics). Typical objective tests used today are the Minnesota Multiphasic Personality Inventory, the Millon Clinical Multiaxial Inventory-III, and the California Psychological Inventory. Common projective tests include the Rorschach inkblot test and the Thematic Apperception Test.
- *Neuropsychological Tests:* Neuropsychological tests consist of specifically designed tasks used to measure psychological functions known to be linked to a particular brain structure or pathway. They are typically used to assess impairment after an injury or illness known to affect neurocognitive functioning, or when used in research, to contast neuropsychological abilities across experimental groups. Examples include the Stroop test, the Bender-Gestalt Test, the Trail Making task, and finger tapping.
- *Clinical Observation:* Clinical psychologists are also trained to gather data by observing behavior. The clinical interview is a vital part of assessment, even when using other formalized tools, which can employ either a structured or unstructered format. Such assessment looks at certain areas, such as general appearance and behavior, mood and affect, perception, comprehension, orientation, insight, memory, and content of communication. One common example of a formal

interview is the mental status examination, which is often used as a screening tool for treatment or further testing.

Diagnostic Impressions

After assessment, clinical psychologists often provide a diagnostic impression. In the U.S., many psychologists use the Diagnostic and Statistical Manual of Mental Disorders (the DSM version IV-TR)-especially when working with an HMO or insurance company-whereas many other countries are more likely to use the International Statistical Classification of Diseases and Related Health Problems. Both assume medical concepts and terms, and state that there are categorical disorders that can be diagnosed by set lists of criteria, which serves psychologists by providing a familiar frame of reference for discussing and understanding the clinical experience and for guiding treatment.

The DSM organizes psychological disorders in five axes:

- *Axis I:* Clinical disorders; Other conditions that may be a focus of clinical attention
    - o Typical disorders include autism, ADHD, dementia, substance abuse, schizophrenia, depression, bipolar, phobias, PTSD, amnesia, anorexia, insomnia, and adjustment disorder
- *Axis II:* Personality disorders (generally rigid and self-defeating traits that are persistent over time and affect day to day living)
    - o Typical disorders include obsessive-compulsive PD, paranoid PD, borderline PD, and narcissistic PD. Mental retardation is also placed on Axis II.
- *Axis III:* medical conditions contributing to the disorder
- *Axis IV:* psychosocial and environmental factors contributing to the disorder
- *Axis V:* Global Assessment of Functioning (on a scale from 100 to 0).

The DSM uses the medical model and views psychological problems in terms of discrete illnesses that can be defined by a minimum set of criteria (such as presenting problems, intensity, behaviors, duration, onset, etc.). While convenient for prescribing medications, there is a growing awareness that this model is not the only way to understand psychological functioning and the various causes of mental distress. As such, there are many debates in the field regarding alternative methods of diagnosing psychological problems.

One such debate is the position of adopting a dimensional model in addition to the current categorical model. Although it is disadvantaged at the moment by a lack of an empirical metric for measuring change or establishing degrees of variation, a dimensional model, according to Jablensky (2005), would have several major advantages, including-addressing quantitative variation and shifts (between various disorders as well as between what is considered normal and pathological); dealing with the presence of multiple conditions; and a more constructive way of looking at otherwise 'sub-threshold' conditions.

Mundt and Backenstrass (2005) describe another variation they call the psychosocial model, which they claim would be far more relevant for the practice of psychotherapy (as opposed to medicine). While the medical model of the DSM is based on assumptions of biology, stability of diagnosis, and objective traits, the psychosocial model is more psychological, intersubjective, and diagnostically flexible over the course of therapy. Like Jablensky, they also suggest that this model be used in conjunction with the medical model, both of which should only be used in the relevant contexts.

British clinical psychologists do not tend to diagnose, but rather use formulation-an individualized map of the difficulties that the patient or client faces, encompassing predisposing, precipitating and perpetuating (maintaining) factors.

Psychotherapy

The central intervention used by clinical psychologists is psychotherapy, which uses a wide range of techniques to change thoughts, feelings, or behaviors in service to enhancing subjective well-being, mental health, and life functioning.

Generally speaking, psychotherapy involves a formal relationship between professional and client-usually an individual, couple, family, or small group-that employs a set of procedures intended to form a therapeutic alliance, explore the nature of psychological problems, and encourage new ways of thinking or behaving.

Although there are literally dozens of recognized therapeutic orientations, their differences can often be categorized on two dimensions: insight vs. action and in-session vs. out-session.

- Insight-emphasis is on gaining greater understanding of the motivations underlying one's thoughts and feelings. (E.g. Psychodynamic therapy)
- Action-focus is on making changes in how one thinks and acts (E.g. Cognitive Behavioral Therapy)
- In-session-interventions center on the hear-and-now interactions between client and therapist (E.g. Humanistic therapy)
- Out-session-although the importance of the relationship is recognized, therapy is geared towards helping the client make changes outside of the session (E.g. Rational Emotive Behavior Therapy)

The methods used are also different in regards to the population being served as well as the context and nature of the problem. Therapy will look very different between, say, a traumatized child, a depressed but high-functioning adult, a group of people recovering from substance dependence, and a ward of the state suffering from terrifying delusions. Other elements that play a critical role in the process of psychotherapy include the environment, culture, age, cognitive functioning, motivation, and duration (i.e. brief or long-term therapy).

The Big Three Perspectives

The field generally recognizes three major perspectives regarding the practice of clinical psychology: Psychodynamic, Cognitive Behavioral, and Humanistic (while a growing debate exists about including the Transpersonal perspective, which recognizes a spiritual dimension in psychological well-being).

Psychodynamic

The Psychodynamic perspective developed out of the Psychoanalysis of Sigmund Freud. The core object of Psychoanalysis is to make the unconscious conscious-to make the client aware of his or her own primal drives (namely those relating to sex and aggression) and the various defenses used to keep them in check. The essential tools of the psychoanalytic process are the use of free association and an examination of the client's transference towards the therapist, defined as the tendency to take unconscious thoughts or emotions about a significant person (e.g. a parent) and apply them to another person who is similar in some way.

Many theorists built upon Freud's fundamental ideas, including Anna Freud, Alfred Adler, Carl Jung, Heinz Hartmann, Karen Horney, Erik Erikson, Ronald Fairbairn, Otto Kernberg, Melanie Klein, Heinz Kohut, Margaret Mahler, David Rapaport, Donald Winnicott, and Harry Stack Sullivan. Major variations on Freudian psychoanalysis include Self Psychology, Ego Psychology, and Object Relations Theory. However, there are still common themes that appear within psychodynamic psychology (as the general perspective is now called, although psychoanalysis in its orthodox form is still practiced), including examination of transference and defenses, an appreciation of the power of the unconscious, and a focus on how early developments in childhood have shaped the client's current psychological state.

Cognitive Behavioral: In the 1950s and 1960s, theorists Albert Ellis and Aaron T. Beck independently began combining the perspectives of Cognitive psychology and Behaviorism to create Cognitive Behavioral Therapy (CBT). Essentially, it is based on the idea that how we think (cognition), how we feel (emotion), and how we act (behavior) all interact together. In this hypothesis, certain thoughts or ways of interpreting the world (called schemas) can cause emotional distress or result in behavioral problems.

The object of CBT is to discover the biased and irrational thinking that leads to emotional problems and to help the client take control over his or her thinking processes in such a way that will lead to increased well-being. There are various approaches along the lines of CBT, such as Rational Emotive Behavior Therapy and Dialectic Behavior Therapy, both of which have been shown to be effective in treating certain conditions, such as depression and phobias.

These theories are informed by a scientific, empirical perspective; clear operationalization of the "problem" or "issue"; an emphasis on measurement (and measurable changes in cognition and behavior); and measurable goal-attainment.

Humanistic

Humanistic psychology was developed in the 1950s in reaction to both behaviorism and psychoanalysis, largely due to the Person-Centered Therapy of Carl Rogers (often referred to as Rogerian Therapy). Rogers believed that a client needed only three things from a clinician to experience therapeutic improvement-congruence, unconditional positive regard, and empathetic understanding. The aim of much humanistic therapy is to give a holistic description of the person. By using Phenomenology, Intersubjectivity and first-person categories, the humanistic psychologist hopes to get a glimpse of the whole person and not just the fragmented parts of the personality. This aspect of holism links up with another aim of humanistic psychology, which is to seek an integration of the whole person, also called self-actualization. According to humanistic thinking, each individual person already has inbuilt potentials and resources that might help them to build a stronger personality and self-concept. The mission of the humanistic psychologist is to help the individual employ these resources via the therapeutic relationship.

Other Major Therapeutic Orientations

There exists literally dozens of recognized schools or orientations of psychotherapy-the list below represents those that have been pivotal in the development of clinical psychology. Although they all have some typical set of techniques practitioners employ, they are generally better known for providing a framework of theory and philosophy that guides a therapist in his or her working with a client.

Systems or Family Therapy

Systems or Family therapy works with couples and families, and emphasizes family relationships as an important factor in psychological health. The central focus tends to be on interpersonal dynamics, especially in terms of how change in one person will affect the entire system. Therapy is therefore conducted with as many significant members of the "system" as possible. In this way, the psychologist will be able to assess and treat client issues within the broader context in which the clients live. There are several different views within Systems therapy depending on the general therapeutic orientation, all of which have somewhat different methodologies and goals. These can include improving communication, establishing healthy roles, creating alternative narratives, and addressing problematic behaviors.

Existential

Existential psychotherapy postulates that people are largely free to choose who we are and how we interpret and interact with the world. It intends to help the client find deeper meaning in life and to accept responsibility for living. As such, it addresses fundamental issues of life-such as death, aloneness, and freedom-and therapy focuses on the development of a client's relationship with these existential issues. The therapist emphasizes the client's ability to be self-aware, freely make choices in the present, establish personal identity and social relationships, create meaning, and cope with the natural anxiety of living.

One influential form of Existential therapy is Gestalt Therapy, primarily founded by Fritz Perls in the mid-20th century. It is well known for techniques designed to increase various kinds of self-awareness-the best-known perhaps being the empty chair technique-which are generally intended to explore resistance to authentic contact, resolve internal conflicts, and help the client complete "unfinished business".

Postmodern

Postmodern psychology says that the experience of reality is a subjective construction built upon language, social context, and history, with no essential truths. Since "mental illness" and "mental health" are not recognized as objective, definable realities, the postmodern psychologist instead sees the goal of therapy as something constructed by the client and therapist. Forms of postmodern psychotherapy include Narrative Therapy, Solution-Focused Therapy, and Coherence Therapy.

Transpersonal

Transpersonal therapy is a relatively new orientation that places a stronger focus on the spiritual facet of human experience. Similar to Existential therapy, it is not a set of techniques so much as a willingness to help a client explore spirituality and/or transcendent states of consciousness. It also is concerned with helping clients achieve his or her highest potential. The orientation is largely informed by Eastern theories on the spiritual and mystical nature of humans, and often focus on enhancing immediate awareness, a fundamental component of meditation.

Important writers in this area include Ken Wilber, Abraham Maslow, Stanislav Grof, John Welwood, and David Brazier.

Integration

In the last couple of decades, there has been a growing movement to integrate the various therapeutic approaches, especially with an increased understanding of cultural, gender, spiritual, and sexual-orientation issues. Clinical psychologists are beginning to look at the various strengths and weaknesses of each orientation while also working with related fields, such as neuroscience, genetics, evolutionary biology, and psychopharmacology. The result is a growing practice of eclecticism, with psychologists learning various systems and the most efficacious methods of therapy with the intent to provide the best solution for any given problem.

Other perspectives

Multiculturalism

Although the theoretical foundations of psychology are rooted in European culture, there is a growing recognition that there exist profound differences between various ethnic and social groups and that systems of psychotherapy need to take those differences into greater consideration. Further, the generations following immigrant migration will have some combination of two or more cultures-with aspects coming from the parents and from the surrounding society-and this process of acculturation can play a strong role in therapy (and might itself be the presenting problem). Culture influences ideas about change, help-seeking, locus of control, authority, and the importance of the individual versus the group, all of which can potentially clash with certain givens in psychotherapeutic theory and practice. As such, more psychologists and training programs are integrating knowledge of various cultural groups in order to inform therapeutic practice in a more culturally sensitive and effective way.

Positive Psychology

Positive psychology is the scientific study of human happiness and well-being, which started to gain momentum in 2000 due to the call of Martin Seligman, then president of the APA. The history of psychology shows that the field has been primarily dedicated to addressing mental illness rather than mental wellness. Applied positive psychology's main focus is on developing positive character strengths and virtues rather than focusing solely on negative problems such as mental disorders.

Feminism

Feminist therapy is an orientation arising from the disparity between the origin of most psychological theories (which have male authors) and the majority of people seeking counseling being female. It focuses on societal, cultural, and political causes and solutions to issues faced in the counseling process. It openly encourages the client to participate in the world in a more social and political way.

5

Educational Psychology

Educational psychology is the study of how humans learn in educational settings, the effectiveness of educational interventions, the psychology of teaching, and the social psychology of schools as organizations. Although the terms "educational psychology" and "school psychology" are often used interchangeably, researchers and theorists are likely to be identified as educational psychologists, whereas practitioners in schools or school-related settings are identified as school psychologists.

Educational psychology is concerned with the processes of educational attainment among the general population and sub-populations such as gifted children and those subject to specific disabilities.

Educational psychology can in part be understood through its relationship with other disciplines. It is informed primarily by psychology, bearing a relationship to that discipline analogous to the relationship between medicine and biology.

Educational psychology in turn informs a wide range of specialities within educational studies, including instructional design, educational technology, curriculum development, organizational learning, special education and classroom management. Educational psychology both draws from and contributes to cognitive science and the learning sciences. In universities, departments of educational psychology are usually housed within faculties of education, possibly accounting for the lack of representation of educational psychology content in introductory psychology textbooks.

Social, Moral and Cognitive Development

To understand the characteristics of learners in childhood, adolescence, adulthood, and old age, educational psychology develops and applies theories of human development. Often cast as stages through which people pass as they mature, developmental theories describe changes in mental abilities (cognition), social roles, moral reasoning, and beliefs about the nature of knowledge.

For example, educational psychologists have researched the instructional applicability of Jean Piaget's theory of development, according to which children mature through four stages of cognitive capability. Piaget hypothesized that children are not capable of abstract logical thought until they are older than about 11 years, and therefore younger children need to be taught using concrete objects and examples. Researchers have found that transitions, such as from concrete to abstract logical thought, do not occur at the same time in all domains. A child may be able to think abstractly about mathematics, but remain limited to concrete thought when reasoning about human relationships. Perhaps Piaget's most enduring contribution is his insight that people actively construct their understanding through a self-regulatory process.

Piaget proposed a developmental theory of moral reasoning in which children progress from a naive understanding of morality based on behavior and outcomes to a more advanced understanding based on intentions. Piaget's views of moral development were elaborated by Kohlberg into a stage theory of moral development. There is evidence that the moral reasoning described in stage theories is not sufficient to account for moral behavior. For example, other factors such as modeling (as described by the social cognitive theory of morality) are required to explain bullying.

Developmental theories are sometimes presented not as shifts between qualitatively different stages, but as gradual increments on separate dimensions. Development of epistemological beliefs (beliefs about knowledge) have been described in terms of gradual changes in people's belief in: certainty and permanence of knowledge, fixedness of ability, and credibility of authorities such as teachers and experts.

People develop more sophisticated beliefs about knowledge as they gain in education and maturity.

Individual Differences and Disabilities

Each person has an individual profile of characteristics, abilities and challenges that result from learning and development. These manifest as individual differences in intelligence, creativity, cognitive style, motivation, and the capacity to process information, communicate, and relate to others. The most prevalent disabilities found among school age children are attention-deficit hyperactivity disorder (ADHD), learning disability, dyslexia, and speech disorder. Less common disabilities include mental retardation, hearing impairment, cerebral palsy, epilepsy, and blindness.

Although theories of intelligence have been discussed by philosophers since Plato, intelligence testing is an invention of educational psychology, and is coincident with the development of that discipline. Continuing debates about the nature of intelligence revolve on whether intelligence can be characterized by a single, scalar factor (Spearman's general intelligence), multiple factors (as in Sternberg's triarchic theory of intelligence and Gardner's theory of multiple intelligences), or whether it can be measured at all. In practice, standardized instruments such as the Stanford-Binet IQ test and the WISC are widely used in economically developed countries to identify children in need of individualized educational treatment. Children classified as gifted are often provided with accelerated or enriched programs. Children with identified deficits may be provided with enhanced education in specific skills such as phonological awareness.

Learning and Cognition

Two fundamental assumptions that underlie formal education systems are that students (a) retain knowledge and skills they acquire in school, and (b) can apply them in situations outside the classroom. But are these assumptions accurate? Research has found that, even when students report not using the knowledge acquired in school, a considerable portion is retained for many years and long term retention is strongly dependent on the initial level of mastery. One study found that university students who took a child development course and attained high grades showed, when tested 10 years later, average retention scores of about 30%, whereas those who obtained moderate or lower grades showed average retention scores of about 20%. There is much less consensus on the crucial question of how much knowledge acquired in school transfers to tasks encountered outside formal educational settings, and how such transfer occurs. Some psychologists claim that research evidence for this type of far transfer is scarce, while others claim there is abundant evidence of far transfer in specific domains. Several perspectives have been established within which the theories of learning used in educational psychology are formed and contested. These include Behaviorism, Cognitivism, Social Cognitivism, and Constructivism. This section summarizes how educational psychology has researched and applied theories within each of these perspectives.

Behavioral Perspective

Applied behavior analysis, a set of techniques based on the behavioral principles of operant conditioning, is effective in a range of educational settings. For example, teachers can improve student behavior by systematically rewarding students who follow classroom rules with praise, stars, or tokens exchangeable for sundry items. Despite the demonstrated efficacy of awards in changing behavior, their use in education has been criticized by proponents of self-determination theory, who claim that praise and other rewards undermine intrinsic motivation. There is evidence that tangible rewards decrease intrinsic motivation in specific situations, such as when the student already has a high level of intrinsic motivation to perform the goal behavior. But the results showing detrimental effects are counterbalanced by evidence that, in other situations, such as when rewards are given for attaining a gradually increasing standard of performance, rewards enhance intrinsic motivation.

Cognitive Perspective

Among current educational psychologists, the cognitive perspective is more widely held than the behavioral perspective perhaps because its flexibly admits causally related mental constructs such as traits, beliefs, memories, motivations and emotions. Cognitive theories posit memory structures that are thought to determine how information is perceived, processed, stored, retrieved and forgotten. Among the memory structures theorized by cognitive psychologists are separate but linked visual and verbal systems described by Allan Paivio's dual coding theory. Educational psychologists have used dual coding theory and cognitive load theory to explain how people learn from multimedia presentations.

The spaced learning effect, a cognitive phenomenon strongly supported by psychological research, has broad applicability within education. For example, students have been found to perform better on a test of knowledge about a text passage when a second reading of the passage is delayed rather than immediate. Educational psychology research has confirmed the applicability to education of other findings from cognitive psychology, such as the benefits of using mnemonics for immediate and delayed retention of information.

Problem solving, regarded by many cognitive psychologists as fundamental to learning, is an important research topic in educational psychology. A student is thought to interpret a problem by assigning it to a schema retrieved from long term memory. When the problem is assigned to the wrong schema, the student's attention is subsequently directed away from features of the problem that are inconsistent with the assigned schema. The critical step of finding a mapping between the problem and a pre-existing schema is often cited as supporting the centrality of analogical thinking to problem solving.

Social Cognitive Perspective

Social cognitive theory is a highly influential fusion of behavioral, cognitive and social elements that was initially developed by educational psychologist Albert Bandura. In its earlier, neo-behavioral incarnation called social learning theory, Bandura emphasized the process of observational learning in which a learner's behavior changes as a result of observing others' behavior and its consequences. The theory identified several factors that determine whether observing a model will affect behavioral or cognitive change. These factors include the learner's developmental status, the perceived prestige and competence of the model, the consequences received by the model, the relevance of the model's behaviors and consequences to the learner's goals, and the learner's self-efficacy. The concept of self-efficacy, which played an important role in later developments of the theory, refers to the learner's belief in his or her ability to perform the modeled behavior.

An experiment by Schunk and Hanson, that studied grade 2 students who had previously experienced difficulty in learning subtraction, illustrates the type of research stimulated by social learning theory. One group of students observed a subtraction demonstration by a teacher and then participated in an instructional program on subtraction. A second group observed other grade 2 students performing the same subtraction procedures and then participated in the same instructional program. The students who observed peer models scored higher on a subtraction post-test and also reported greater confidence in their subtraction ability. The results were interpreted as supporting the hypothesis that perceived similarity of the model to the learner increases self-efficacy, leading to more effective learning of modeled behavior. It is supposed that peer modeling is particularly effective for students who have low self-efficacy.

Over the last decade, much research activity in educational psychology has focused on developing theories of self-regulated learning (SRL) and metacognition. These theories work from the central premise that effective learners are active agents who construct knowledge by setting goals, analysing tasks, planning strategies and monitoring their understanding. Research has indicated that learners' who are better at goal setting and self-monitoring tend to have greater intrinsic task interest and self-efficacy; and that teaching learning strategies can increase academic achievement.

Constructivist Perspective

Constructivism refers to a category of learning theories in which emphasis is placed on the agency and prior knowledge of the learner, and often on the social and cultural determinants of the learning process. Educational psychologists distinguish individual (or psychological) constructivism, identified with

Piaget's learning theory, from social constructivism. A dominant influence on the latter type is Lev Vygotsky's work on sociocultural learning, describing how interactions with adults, more capable peers, and cognitive tools are internalized to form mental constructs. Elaborating on Vygotsky's theory, Jerome Bruner and other educational psychologists developed the important concept of instructional scaffolding, in which the social or information environment offers supports for learning that are gradually withdrawn as they become internalized.

Vygotsky's version of constructivist theory has led to the view that behavior, skills, attitudes and beliefs are inherently situated, that is, bound to a specific sociocultural setting. According to this view, the learner is enculturated through social interactions within a community of practice. The social constructivist view of learning has spawned approaches to teaching and learning such as cognitive apprenticeship, in which the tacit components of a complex skill are made explicit through conversational interactions occurring between expert and novice in the setting in which the skill is embedded.

Motivation

Motivation is an internal state that activates, guides and sustains behavior. Educational psychology research on motivation is concerned with the volition or will that students bring to a task, their level of interest and intrinsic motivation, the personally held goals that guide their behavior, and their belief about the causes of their success or failure.

A form of attribution theory developed by Bernard Weiner describes how students' beliefs about the causes of academic success or failure affect their emotions and motivations. For example, when students attribute failure to lack of ability, and ability is perceived as uncontrollable, they experience the emotions of shame and embarrassment and consequently decrease effort and show poorer performance. In contrast, when students attribute failure to lack of effort, and effort is perceived as controllable, they experience the emotion of guilt and consequently increase effort and show improved performance.

Motivational theories also explain how learners' goals affect the way that they engage with academic tasks. Those who have mastery goals strive to increase their ability and knowledge. Those who have performance approach goals strive for high grades and seek opportunities to demonstrate their abilities. Those who have performance avoidance goals are driven by fear of failure and avoid situations where their abilities are exposed. Research has found that mastery goals are associated with many positive outcomes such as persistence in the face of failure, preference for challenging tasks, creativity and intrinsic motivation. Performance avoidance goals are associated with negative outcomes such as poor concentration while studying, disorganized studying, less self-regulation, shallow information processing and test anxiety. Performance approach goals are associated with positive outcomes, and some negative outcomes such as an unwillingness to seek help and shallow information processing.

Research Methodology

The research methods used in educational psychology tend to be drawn from psychology and other social sciences. There is also a history of significant methodological innovation by educational psychologists, and psychologists investigating educational problems. Research methods address problems in both research design and data analysis. Research design informs the planning of experiments and observational studies to ensure that their results have internal, external and ecological validity. Data analysis encompasses methods for processing both quantitive (numerical) and qualitative (non-numerical) research data. Although, historically, the use of quantitative methods was often considered an essential mark of scholarship, modern educational psychology research uses both quantitative and qualitative methods.

Quantitative Methods

Perhaps first among the important methodological innovations of educational psychology was the development and application of factor analysis by Charles Spearman. Factor analysis is mentioned here as one example of the many multivariate statistical methods used by educational psychologists. Factor analysis is used to summarize relationships among a large set of variables or test questions, develop theories about mental constructs such as self-efficacy or anxiety, and assess the reliability and validity of test scores. Over one hundred years after its introduction by Spearman, factor analysis has become a research staple figuring prominently in educational psychology journals.

Because educational assessment is fundamental to most quantitative research in the field, educational psychologists have made significant contributions to the field of psychometrics. For example, alpha, the widely used measure of test reliability was developed by educational psychologist Lee Cronbach. The reliability of assessments are routinely reported in quantitative educational research. Although, originally, educational measurement methods were built on classical test theory, item response theory and Rasch models are now used extensively in educational measurement worldwide. These models afford advantages over classical test theory, including the capacity to produce standard errors of measurement for each score or pattern of scores on assessments and the capacity to handle missing responses.

Meta-analysis, the combination of individual research results to produce a quantitative literature review, is another methodological innovation with a close association to educational psychology. In a meta-analysis, effect sizes that represent, for example, the differences between treatment groups in a set of similar experiments, are averaged to obtain a single aggregate value representing the best estimate of the effect of treatment. Several decades after Pearson's work with early versions of meta-analysis, Glass published the first application of modern meta-analytic techniques and triggered their broad application across the social and biomedical sciences. Today, meta-analysis is among the most common types of literature review found in educational psychology research.

Qualitative Methods

Qualitative methods are used in educational studies whose purpose is to describe events, processes and situations of theoretical significance. The qualitative methods used in educational psychology often derive from anthropology, sociology or sociolinguistics. For example, the anthropological method of ethnography has been used to describe teaching and learning in classrooms. In studies of this type, the researcher may gather detailed field notes as a participant observer or passive observer. Later, the notes and other data may be categorized and interpreted by methods such as grounded theory. Triangulation, the practice of cross-checking findings with multiple data sources, is highly valued in qualitative research.

Case studies are forms of qualitative research focusing on a single person, organization, event, or other entity. In one case study, researchers conducted a 150-minute, semi-structured interview with a 20-year old woman who had a history of suicidal thinking between the ages of 14 to 18. They analyzed an audio-recording of the interview to understand the roles of cognitive development, identity formation and social attachment in ending her suicidal thinking.

Qualitative analysis is most often applied to verbal data from sources such as conversations, interviews, focus groups, and personal journals. Qualitative methods are thus, typically, approaches to gathering, processing and reporting verbal data. One of the most commonly used methods for qualitative research in educational psychology is protocol analysis. In this method the research participant is asked to think aloud while performing a task, such as solving a math problem. In protocol analysis the verbal data is thought to indicate which information the subject is attending to, but is explicitly not interpreted as an explanation or justification for behavior.

In contrast, the method of verbal analysis does admit learners' explanations as a way to reveal their mental model or misconceptions (e.g., of the laws of motion). The most fundamental operations in both protocol and verbal analysis are segmenting (isolating) and categorizing sections of verbal data. Conversation analysis and discourse analysis, sociolinguistic methods that focus more specifically on the structure of conversational interchange (e.g., between a teacher and student), have been used to assess the process of conceptual change in science learning. Qualitative methods are also used to analyse information in a variety of media, such as students' drawings and concept maps, video-recorded interactions, and computer log records.

Applications in Instructional Design and Technology

Instructional design, the systematic design of materials, activities and interactive environments for learning, is broadly informed by educational psychology theories and research. For example, in defining learning goals or objectives, instructional designers often use a taxonomy of educational objectives created by Benjamin Bloom and colleagues.

Bloom also researched mastery learning, an instructional strategy in which learners only advance to a new learning objective after they have mastered its prerequisite objectives. Bloom discovered that a combination of mastery learning with one-to-one tutoring is highly effective, producing learning outcomes far exceeding those normally achieved in classroom instruction. Gagné, another psychologist, had earlier developed an influential method of task analysis in which a terminal learning goal is expanded into a hierarchy of learning objectives connected by prerequisite relations.

Applications in Teaching

Research on classroom management and pedagogy is conducted to guide teaching practice and form a foundation for teacher education programs. The goals of classroom management are to create an environment conducive to learning and to develop students' self-management skills. More specifically, classroom management strives to create positive teacher-student and peer relationships, manage student groups to sustain on-task behavior, and use counselling and other psychological methods to aid students who present persistent psychosocial problems.

History

Educational psychology cannot claim priority in the systematic analysis of educational processes. Philosophers of education such as Democritus, Quintilian, Vives and Comenius, had examined, classified and judged the methods of education centuries before the beginnings of psychology in the late 1800s. Instead, aspirations of the new discipline rested on the application of the scientific methods of observation and experimentation to educational problems. Even in the earliest years of the discipline, educational psychologists recognized the limitations of this new approach. In his famous series of lectures Talks to Teachers on Psychology, published in 1899 and now regarded as the first educational psychology textbook, the pioneering American psychologist William James commented that:

Psychology is a science, and teaching is an art; and sciences never generate arts directly out of themselves. An intermediate inventive mind must make that application, by using its originality.

According to Berliner educational psychology theorists' attitude to the world of educational practice has shifted from initial interest to disdain, and eventually to respect.

In 1912, Thorndike, who developed the theory of instrumental conditioning, presaged later work on programmed instruction, mastery learning and computer-based learning:

If, by a miracle of mechanical ingenuity, a book could be so arranged that only to him who had done what was directed on page one would page two become visible, and so on, much that now requires personal instruction could be managed by print.

Influential Educational Psychologists and Theorists

The following persons were selected and featured in a recent biographical history of educational psychology as having made significant contributions to the field:

- Albert Bandura 1925-
- Alfred Binet 1857-1911
- Benjamin Bloom 1913-1999
- Ann Brown 1943-1999
- Jerome Bruner 1915-
- Lee Cronbach 1916-2001
- John Dewey 1859-1952
- Nathaniel Gage 1917-
- Robert Gagné 1916-2002

- William James 1842-1910
- Maria Montessori 1870-1952
- Jean Piaget 1896-1980
- Herbert Simon 1916-2001
- Burrhus Frederic Skinner 1904-1990
- Charles Spearman 1863-1945
- Lewis Terman 1877-1956
- Edward L. Thorndike 1874-1949
- Lev Semenovich Vygotsky 1896-1934

Careers in Educational Psychology

Education and Training

A person may be considered an educational psychologist if he or she has completed a graduate degree in educational psychology or a closely related field. Universities establish educational psychology graduate programs in either psychology departments or, more commonly, faculties of education.

Educational psychologists are usually trained at the doctoral level (PhD) and typically work in university settings, where they carry out research on topics related to the cognitive and social processes involved in learning and education.

Educational psychologist may also work as consultants in designing and creating educational materials and classroom programs designed to promote learning. Within the last few years, with the growing popularity of online education, a new movement has been for educational psychologists to be involved with studying and designing online courses and learning strategies.

A similar field to educational psychology is school psychology. Psychologists who work in a k-12 school setting are usually trained at either the masters or doctoral (PhD or EdD) level. In addition to conducting assessments, school psychologists provide services such as academic and behavioral intervention, counseling, teacher consultation, and crisis intervention.

Employment Outlook

Employment for psychologists in the United States is expected to grow faster than most occupations through the year 2014, with anticipated growth of 18-26%. One in four psychologist are employed in educational settings. In the United States, the median salary for psychologists in primary and secondary schools is $58,360 as of May 2004.

In recent decades the participation of women as professional researchers in North American educational psychology has risen dramatically. The percentage of female authors of peer-reviewed journal articles doubled from 1976 (24%) to 1995 (51%), and has since remained constant. Female membership on educational psychology journal editorial boards increased from 17% in 1976 to 47% in 2004. Over the same period, the proportion of chief editor positions held by women increased from 22% to 70%.

6

Forensic and Health Psychology

Forensic psychology is the application of psychological principles and knowledge to various legal activities including child custody disputes, child abuse of an emotional, physical and sexual nature, assessing one's personal capacity to manage one's affairs, matters of competency to stand trial, criminal responsibility and personal injury and advising judges in matters relating to sentencing regarding various mitigants and the actuarial assessment of future risk.

One of the earliest examples of a psychologist acting as an expert witness in a court of law was in 1896 when Albert von Schrenck-Notzing testified at the trial of a man accused of murdering three women. Drawing on research into memory and suggestibility he argued that pre-trial publicity meant that witnesses could not distinguish between what they actually saw and what had been reported in the press.

In the civil law arena, forensic psychologists often provide assessments of whether someone has been harmed by some event. For example, in a wrongful death suit, a psychologist might offer opinions as to whether a plaintiff suffered emotional trauma in response to the death of a loved one. They might also assess the emotional injuries suffered by someone who has been injured in an accident or who witnessed a traumatic event. Psychologists are often called upon in sexual harassment suits to describe the impact of the harassment on the purported victim. In this arena, the forensic psychologist might be required to provide treatment recommendations or to analyze the specific treatment needs of an individual, and might be asked to determine the potential cost of such treatment.

In the arena of workers' compensation law, a forensic psychologist might be called upon to describe how workplace stress factors impacted the psychological functioning of a claimant, or to determine whether the purported work place stress had any effect on the worker at all. As in the more general civil law context, the forensic psychologist might be asked to determine treatment needs and treatment plans.

In the family law arena, forensic psychologists are often called upon to assess the "best interests" of children whose parents are divorcing. Commonly, this involves making recommendations to a Court with respect to child custody arrangements. Child custody mediation is another role that forensic psychologists undertake in the family law arena-serving as a mediator between divorced parents who remain in dispute about the needs and interests of their children. In some jurisdictions, forensic psychologists are appointed as "special masters" by the Court, and are charged with making both recommendations and orders for the care of children in disputed custody situations.

Forensic psychologists are perhaps most commonly recognized for their involvement in the processing a crime scene. Psychologists provide Courts with analysis relevant to questions of criminal insanity and trial competence. They help courts decide whether sex offenders are likely to re-offend or whether they are dangerous. They provide information and recommendations necessary for sentencing purposes, grants of probation, and the formulation of conditions of parole. Forensic psychologists are routinely called upon in death penalty cases to provide analysis of the intentions, motivations and personality characteristics of the accused. In the juvenile courts, they often are asked to help determine whether a youthful offender can be rehabilitated. They assist prosecutors, defenders, and law enforcement investigators in understanding a range of normal and criminal behaviors, sometimes serving as "criminal profilers".

HEALTH PSYCHOLOGY

Health psychology is a relatively new field which is evolving and developing as one of the main areas of applied psychology.

Although its early beginnings can be traced to the kindred field of clinical psychology, four different approaches to health psychology have been defined: clinical, public health, community and critical health psychology (Marks, Murray et al., 2005).

Many researchers believe that physical health may be influenced by psychology through a variety of direct and indirect means. There is some evidence that certain negative mental states (such as depression and anxiety) can directly affect physical immunity through production of stress hormones, such as the catecholamines and glucocorticoids.

Although this research is widely debated, there is also some indication that negative psychological states may lead to faster disease progression in certain diseases (such as HIV and heart disease) through these direct biological mechanisms. Negative emotional states may also indirectly affect disease processes through their influence on health behaviors. For example, depression has been related to many risk factors for poor health including overeating, smoking, physical inactivity, and poor medication compliance.

Health psychology is the "...use of psychological principles to promote health and to prevent illness" (Taylor 1990); it is also part of clinical treatment for established illness. This approach considers the biological, cognitive, behavioral, emotional, social, psychosomatic and environmental factors as they relate to health, illness and health care at the level of individuals. This approach has adopted what it calls the biopsychosocial model.

- Clinical health psychology (ClHP) adopts a definition reflecting the fact that the field was originally a branch of clinical psychology; it is also a major contributor to the field of behavioral medicine within psychiatry. Clinical practice includes the techniques of education, behavioral change, and psychotherapy; with additional training, a clinical health psychologist can become a medical psychologist in certain countries and, consequently, prescribe medication.
- Organizational health psychology (OHP) is the use of health psychology knowledge and techniques as applied

to health and illness in the workplace, at the individual and group levels.

- Public health psychology (PHP) tries to determine causation between psychological factors and health for the population as a whole, and present this information to educators, policy makers, and health care practitioners for the promotion of better public health. It gives prominence to population-level health outcomes and interventions. It is allied to other disciplines in the fields of global and public health including epidemiology, nutrition, genetics and statistics. Interventions are determined and "targeted" using population health statistics which evaluate health needs of various population groups as perceived by health authorities and policy makers. These interventions are "top-down" and tend to have varying levels of effectiveness across diverse population groups.
- Community health psychology (CoHP) approach tries to understand what is happening at a local level. Interventions are generated collaboratively with coalitions of stakeholders and are aimed to facilitate community empowerment to improve the physical and/or mental health of local people outside of any formal involvement of the health care system.
- Critical health psychology (CrHP) is concerned with the distribution of power and the impact of power differentials on health experience and behaviour, health care systems and health policy. It prioritises social justice and the universal right to health of peoples of all races, genders, ages, and socioeconomic positions. A major concern is health inequalities. The CrHPist is an agent of change, not simply an analyst or cataloguer. The field was developed through the foundation of the Journal of Health Psychology and the International Society of Critical Health Psychology.

Health psychology is both a theoretical and applied field. Many different methods are employed including questionnaires, interviews, controlled studies, and actions designed to bring about change using "action research". Health psychologists conduct health interviews with clients that aim to construct a more holistic picture of each person's health, one that includes their genes, religious beliefs, social supports, living conditions, emotional state, and beliefs of health, etc. They use this information to work alongside a person's physicians and therapists to develop a treatment tailored for individual needs or to develop greater empowerment among the community's members so that the community is able to strengthen and sustain its own quality of life.

Objectives of Health Psychology

Understand behavioral Factors

Health psychologists seek to identify the behaviors and experiences that promote health, lead to illness, influence the effectiveness of health care, and recommend improvements to health policy in their light. There are many examples of this. Smoking, diet, and regular exercise all contribute to the formation of disease. There are some minor associations between illness and individual characteristics such as personality. For example, it is claimed that individuals with thrill seeking personalities are more likely to drive fast, making them more likely to injure themselves in car accidents. And also that people distrustful of physicians will not get regular checkups. However there are contextual factors in the form of political, economic, cultural, community, social and lifestyle factors that have a more major influence on the health of particular individulas, although they will not generally recognise these.

Factors that lead to the behaviors that cause illness are of interest because they help psychologists to predict who is most susceptible to illness and why. There are many contributing factors that help determine our behaviors, and all of the dimensions of the biopsychosocial model can be applied to understand these interconnectons. Biologically, physical addiction plays an important role in smoking cessation. As does the psychological dependency on tobacco brought about largely by seductive advertising and other forms of tobacco promotion. Psychologically, people with high stress jobs are more likely to develop cardiovascular disease. Socially, people with low incomes have less access to health resources and screening processes.

They also are exposed to greater environmental toxigens, lower levels of education, poorer housing, less healthy foods, higher smoking prevalence, and many other toxic socially determined living conditions.

Health psychologists also aim to change health behaviors for the dual purpose of helping people stay healthy and helping patients adhere to disease treatment regimens. Cognitive behavioral therapy and behavior modification are techniques often used for this purpose.

Prevent Illness

Psychologists work towards promoting health through behavioral change, as mentioned above, but they prevent illness in other ways as well. Practitioners emphasize education as a large part of illness prevention, as many people do not recognize the risk to illness present in their lives. Or they are unable to implement the knowledge that they have owing to the pressures of their everyday existence. A common example of this is anti-smoking campaigns. Those least able to afford tobacco products consume them the most. It is a method for controlling emotional states, the daily experiences of stress that characterise the lives of deprived and vulnerable people. Health psychologists also aim at educating health professionals like physicians and nurses in communicating with patients in a way that highlights the psychosocial barriers to understanding, memorising and implementing effective strategies for reducing risk factors and making behaviour changes and that will help them to develop appropriate means for communicating information relevant to diagnosis, treatment and prognosis (Ogden, 2004).

Explore the Effects of Disease

There is much to know about how disease affects our mental well being. When illness or accidents befall a person, their entire life is affected. A psychologically healthy individual who gets severely injured, say, now has many different practical issues to contend with that will in turn affect their psychological wellbeing. Who will take care of them while they recover? If they can't work, how will they pay for bills or care for dependents? If this person sees themselves as being self-reliant, how do they handle this new identity? What if they cannot pursue their usual hobbies and interests? All of these possibilities can affect a person's relationships, self-esteem, stress level, happiness and belief system. Many of these issues are economic and social in nature rather than psychological but they have a direct impact on psychological wellbeing.

This important field of study considers how those with terminal illnesses can lead a better life. When there is little hope of recovery, health psychology therapists can improve the quality of life of the patient by helping them recover their mental well-being.

Critical Analysis of Health Policy

Critical health psychologists are exploring how health policy can impact on inequities, inequalities and social injustice. This expands the scope of health psychology beyond the level of individual health to an examination of the social and economic determinants of health experience both within and between regions and nations. The individualism of mainstream health psychology has been critiqued and deconstructed by critical health psychologists and newer qualitative methods and frameworks for investigating health experience and behaviour are advocated (Marks, Murray et al., 2005).

Improving Doctor-patient Communication

Health psychologists attempt to aid the process of communication between doctors and patients during medical consultations. There are many problems in this process, particularly the use of jargon by doctors (using long, complex, usually medicine-specific words often not understood by the patient). Researchers such as Boyle and McKinlay have investigated this process and discovered not only that patient understanding is low (40% of women on a maternity ward understood 13 medical terms that they were given), but that even doctors do not expect their patients to understand their jargon! One main area of research on this topic involves 'doctor centred' or 'patient centred' consultations. Doctor centred consultations are generally directive, with the patient answering questions and taking little part in decision-making. Although this style is preferred by elderly people and others, many people dislike the sense of hierarchy or ignorance that it inspires. They prefer the patient centred consultations, which focus on the patient's needs, involve the doctor listening to the patient completely before making a decision, and involving the patient in the process of choosing treatment and finding a diagnosis.

Adherence to Medical Advice

Getting people to obey medical advice and adhere to their treatment regimens is a difficult task for health psychologists. Mostly, people forget to take their pills or find their side effects too difficult to cope with (rational non-adherence), but failing to take prescribed medication costs the Health Service millions of pounds per year, as well as wasting millions of usable, viable medicines that could otherwise be used to help other people. Estimated adherence rates are very hard to measure due to the issues raised by the various techniques available. Sarafino estimates adherence to be very low, however, he judges adherence to be following every part of the treatment programme, whereas Taylor encourages the tailoring of the programme to one's personal routine.

Ways of Measuring Adherence

- Counting the number of pills in the medicine bottle-although this has problems with privacy and seems patronising or showing lack of trust in patients
- Using self report measures-although participants may fail to return the self report or lie about their adherence
- Asking doctor or health worker-although this presents problems on doctor-patient confidentiality
- Using 'Trackcap' bottles, which measure the number of times the bottle is opened-however, this either raises problems of informed consent or, if informed consent is gained, demand characteristics.

RELATIONSHIP COUNSELING

Relationship counseling is the process of counseling the parties of a relationship in an effort to recognize and to better manage or reconcile troublesome differences. The relationship involved may be between members of a family, couples, employees or employers in a workplace, or between a professional and a client.

History

Relationship counseling as a discrete, professional service is a recent phenomenon. Until the late 20th century, the work of relationship counseling was informally fulfilled by close friends, family members, or local religious leaders. Psychologists and psychotherapists have historically dealt primarily with individual psychological problems. In many less technologically advanced cultures around the world today, the institution of family, the village or group elders fulfill the work of relationship counseling.

With increasing modernization or westernization in many parts of the world and the continuous shift towards isolated nuclear families, the old support structures are no longer there and the need for relationship counseling is greater than ever. In western society the trend is towards trained relationship counselors; these are often volunteers who wish to help others, and are trained by either the Government or social service institutions to help those who are in need of counseling. Many communities and government departments have their own team of trained voluntary or professional relationship counselors. Similar services are operated by many universities and colleges, often staffed by volunteers from among the student peer group. Some large companies maintain a full-time professional counseling staff to facilitate smoother interactions between corporate employees, and prevent personal difficulties affecting work performance.

Methodology

Before the relationships between the individuals can begin to be understood, it is important for all to recognize and acknowledge that everyone involved has a unique personality, and background. Sometimes the individuals in the relationship adhere to different value systems. Institutional and societal variables (like the social, religious, group and other collective factors) which shape a person's nature, and behavior must be recognized. A tenet of "relationship counseling" is that:

It is intrinsically beneficial for all the participants to interact with each other and with society at large with the least conflict possible.

Occasionally the relationships get 'strained', which means that they are not functioning at the optimum extent. There are many possible reasons for this, including ego, arrogance, jealousy, anger, greed etc. Often it is an interaction between two or more factors, and frequently it is not just one of the people who are involved that exhibits such traits.

Some say the only viable solution to the problem of setting these relationships back on track is to reorient the individuals' perceptions-how one looks at or responds to situations. This implies that they make some fundamental changes in their attitudes-much easier said than done. The next step is to adopted conscious structural changes to their inter-personal relationships.

The duty and function of a relationship counselor is to listen, understand and facilitate a better understanding between those involved. The basic principles involved are

- non-judgment on any of the issues or incidents narrated to them as counselor.
- Confidentiality of the persons being given the counseling.

A successful counselor is someone who has a mature and balanced state of mind and disposition, who can place themselves in the shoes of those they are counseling, and the ability to respect their opinions, thoughts, feelings and (more importantly) emotions.

After evaluating the story as it is narrated, a realistic, practical solution can be developed; individually at first if this is beneficial, and then jointly to encourage the participants to give their best efforts at reorienting their relationship with each other. It has to be remembered that the change in situations like financial state, physical health, and the influence of other family members can have a profound influence on the conduct, responses and actions of the individuals.

Couples Counseling

Couples counseling involves a couple meeting with the psychologist, social worker or other type of mental health professional for counseling to address the dysfunction in their marriage or other type of relationship.

It is commonly observed that many couples who seek counseling dissolve their relationship despite their clear intention to avoid this extreme emotional trauma and expense. The most successful work in counseling involves the study of couples whose relationship has been restored and who have found a counselor capable of creating an appropriate relationship education milieu.

There are many studies indicating that troubled couples have great difficulties with a "value neutral" approach, when they are diligently seeking to resolve difficulties in their marriage. The neutrality is seen as therapeutic vacuum. The dearth of table of contents entries pertaining to the restoration of marital health in psychiatric, psychological, social work, and counseling text books and journal articles indicates a specific professional de-emphasis on relationship education, restoration, forgiveness, and healing. A general denial of couples goal to restore the very positive values and experiences that characterized their relationship in the past may represent a significant part in the negative assessment of the Consumer Reports respondents mentioned below.

Occasionally, divorced couples use mediation to resolve the matters of custody, spousal support and the division of property. The use of the same professional to re-build a marriage or to end it might appear incongruous and troublesome to the majority of couples who are seeking help at a very distressed time in their lives.

Many individuals refuse to seek counseling because of the feeling that they are admitting that their marriage has failed or that locating a counselor capable of restoration can be accomplished readily. However, many couples in minimally-distressed marriages seek counseling to resolve difficult concerns, to confront problems in the context of couples therapy or to find a neutral location to improve their relationship. The most successful marriage counselors may meet with the partners separately before meeting with them together, or may even have individual counselors who meet with the partners and then have a group session with all the counselors and the partners. In common practice, though, the latter is rare, and it is over-wrought with concerns about confidentiality and cost management. Having one assistant to see each party separately, to assess strengths and weakness, and then to reconvene with both to discuss what each can do for the other to rebuild is a most helpful technique.

The internet has added new dimensions to traditional face to face counseling. It is now possible to engage in counseling sessions with therapists in other states or even other countries via web cams, email and the telephone. Websites can easily outline the service intervention approach and provide free and widely accessible readings that may challenge inappropriate and resentful assumptions about a partner, all the while re-establishing the common ground and reinforcing the strengths that brought the couple together, initially. This step, alone can be very helpful.

Most studies purporting to show counseling effectiveness do not provide parallel study of placebo groups and do not track improvement that may come with the mere passage of time, in the absence of treatment. A 1995 study by Consumer Reports, with a sample size of over 15,000, while indicating some value in counseling by the North American consumers of these services, indicates substantial dissatisfaction with marriage counseling. Marriage counseling was seen as the least effective intervention offered. (Many counseling agencies will not provide longer term studies of their work, and have no means of inquiry as to the marital status of couples in subsequent years.)

The type of counseling that was reported as the most valuable by consumers was Alcoholic Anonymous, a lay, peer support. The work of the Coalition for Marriage, Family, and Couples Education, also reports substantial improvement for couples by means outside of traditional, pathology-based psychotherapy.

Counseling or therapy that is reimbursed by health insurance in the United States, requires a diagnosis of mental illness. Psychotherapy methods rarely involve instruction to the couple, or to the husband or wife regarding specific methods to correct behaviors which have led to deterioration in the marriage. This void is being met by the marriage education movement which dates its beginnings to 1995.

This movement, while not seeking to supplant therapy or counseling, sees nearly all individuals and couples, including unmarried, dating couples, as capable of learning improved relationship skills, from a variety of means, including self-study. Intensive follow-up at the University of Denver has shown repeatedly, that educational efforts, when presented by lay persons, can often be more efficacious than therapy given by pathology-bound psychotherapists.

7

School Psychology

School Psychology is a field that applies principles of clinical psychology and educational psychology to the diagnosis and treatment of students' behavioral and learning problems. School psychologists are educated in child and adolescent development, learning theories, psychoeducational assessment, personality theories, therapeutic interventions, identification of learning disabilities, and the ethical, legal and administrative codes of their profession.

Historical Highlights

School psychology began mainly through the testing movement, in the late 1800s, especially from people like Alfred Binet. Binet's work resembled school psychology of today, because he developed the first IQ test to screen children who would not benefit from regular education. Binet's test was brought to the United States in the early 1900s, and was standardized in 1916 by Lewis Terman of Stanford University. Today it is known has the Stanford-Binet test. In 1975, the Education of All Handicapped Children Act (EHA) mandated the free and appropriate education of all individuals from 3-21 years of age. This act requires that all children should attend school, including children who often would not receive any advantages of public education in the past due to their disability. Its principles were reaffirmed and strengthened in the Individuals with Disabilities Education Act (IDEA), and Individuals with Disabilities Education Improvement Act (IDEIA). Hence, in the U.S. the profession of school psychology flourished as these children needed additional support to be included in the regular school setting. This combined with years of litigation allowed the profession to flourish (Graduate School of Education, 2003).

Theoretical Framework and Services

According to Division 16 (Division of School Psychology), of the American Psychological Association (APA) school psychologists operate according to a scientific framework. They work to promote effectiveness and efficiency in the field. School psychologists conduct psychological assessments, provide brief interventions, and develop or help develop prevention programs. Additionally, they evaluate services with special focus on developmental processes of children within the school system, and other systems, such as families. School psychologists consult with teachers, parents, and school personnel about learning and behavioral problems. They may teach lessons on parenting skills (like school counselors), learning strategies, and other skills related to school health. In addition, they often have to explain test results to parents and students. They also may do some counseling (State Board of Education 2003; National Clearinghouse for Professions in Special Education, n.d.).

Education and Certification

The APA only accredits doctoral programs in school psychology. Its standards describe how the program should be structured, but not specific courses to be offered (Committee on Accreditation, 2002). Likewise, the U.S. National Association of School Psychologists (NASP) describes how the program should be developed and implemented. In the U.S., a Specialist degree is usually required for an individual to work as a school psychologist, although a few states still credential school psychologists with a Master's degree or Certificate of Advanced Graduate Studies (CAGS). In addition, NASP provides a national credential (NCSP) for those who have the equivalent of a Master's degree plus 30 graduate semester hours, a 1200-hour supervised internship, and have a passing score on the National School Psychology Examination (National Clearinghouse for Professions in Special Education, n.d.).

Accredited programs require courses/practica/internships to ultimately cover the following domains: data-based decision-making and accountability; consultation and collaboration; effective instruction and development of cognitive/academic skills; socialization and development of life skills; student diversity in development and learning; school and systems organization, policy development, and climate; prevention, crisis intervention, and mental health; home/school/community collaboration; research and program evaluation; school psychology practice and development; and information technology. Ph.D. training programs differ from Specialist programs in that they require students to take courses within a chosen area of academic specialty, complete a dissertation and comprehensive examination, and complete a qualified 12-month (as opposed to 9-month) internship. Ph.D. programs may also require students to involve themselves in more advanced research endeavors within the department. While the Ed.S. and SSP are the most common Specialist degrees, some schools instead offer a Psy.S. (Specialist in Psychology) in school psychology, which is a post-Master's specialist degree.

Since school psychologists are so influential within the school system and frequently consulted to solve problems, practitioners should be able to collaborate with other members of the educational community and confidently make decisions based on empirical research.

Behaviorism

Behaviorism is an approach to psychology based on the proposition that behavior can be studied and explained scientifically without recourse to internal mental states. A similar approach to political science may be found in Behavioralism. The behaviorist school of thought ran concurrent with the psychoanalysis movement in psychology in the 20th century. Its main influences were Ivan Pavlov, who investigated classical conditioning, John B. Watson who rejected introspective methods and sought to restrict psychology to experimental methods, and B.F. Skinner who conducted research on operant conditioning.

Approaches

Within that broad approach, there are different emphases. Some behaviorists argue simply that the observation of behavior is the best or most convenient way of investigating psychological and mental processes. Others believe that it is in fact the only way of investigating such processes, while still others argue that behavior itself is the only appropriate subject of psychology, and that common psychological terms (belief, objectives, etc.) have no referents and/or only refer to behavior. Those taking this point of view sometimes refer to their field of study as behavior analysis or behavioral science rather than psychology.

Albert Bandura's social cognitive approach grew out of this movement as well. Bandura's approach stresses mental (cognitive) processes in addition to observable behavior, focusing on not only the impact of the environment but also "observation, imitation, and thought processes".

There is no classification generally agreed upon, but some would say it as true, and some would add to or modify this list.

- *Classical:* The behaviorism of Watson; the objective study of behavior; no mental life, no internal states; thought is covert speech.
- *Methodological:* The objective study of third-person behavior; the data of psychology must be inter-subjectively verifiable; no theoretical prescriptions. It has been absorbed into general experimental and cognitive psychology.
- *Radical:* Skinnerian behaviorism; is considered radical since it expands behavioral principles to processes within the organism; in contrast to methodological behaviorism; not mechanistic or reductionist; hypothetical (mentalistic) internal states are not considered causes of behavior, phenomena must be observable at least to the individual experiencing them.
- *Logical:* Established by Oxford philosopher Gilbert Ryle in his book The Concept of Mind (1949).
- *Teleological:* Post-Skinnerian, purposive, close to microeconomics.
- *Theoretical:* Post-Skinnerian, accepts observable internal states ("within the skin" once meant "unobservable", but with modern technology we are not so constrained);

dynamic, but eclectic in choice of theoretical structures, emphasizes parsimony.

Biological: Post-Skinnerian, centered on perceptual and motor modules of behavior, theory of behavior systems.

Interbehaviorism: Founded by J. R. Kantor before Skinner's writings and currently worked by L. Hayes; E, Ribes; and S. Bijou. Centered in the interbehavior of organisms, field theory of behavior; emphasis on human behavior.

Two popular subtypes are Neo-: Hullian and post-Hullian, theoretical, group data, not dynamic, physiological, and Purposive: Tolman's behavioristic anticipation of cognitive psychology.

Watson's Argument

Watson argued in his book Psychology from the Standpoint of a Behaviorist (1919) for the value of a psychology which concerned itself with behavior in and of itself, not as a method of studying consciousness. This was a substantial break from the structuralist psychology of the time, which used the method of introspection and considered the study of behavior valueless. Watson, in contrast, studied the adjustment of organisms to their environments, more specifically the particular stimuli leading organisms to make their responses. Most of Watson's work was comparative, i.e., he studied the behavior of animals. Watson's approach was much influenced by the work of Russian physiologist Ivan Pavlov, who had stumbled upon the phenomenon of classical conditioning (learned reflexes) in his study of the digestive system of the dog, and subsequently investigated the phenomena in detail. Watson's approach emphasized physiology and the role of stimuli in producing conditioned responses-assimilating most or all function to reflex. For this reason, Watson may be described as an S-R (stimulus-response) psychologist.

Methodological Behaviorism

Watson's Behaviorist theory persuaded most academic researchers in experimental psychology of the importance of studying behavior. In the field of comparative psychology in particular, it was consistent with the warning note that had been struck by Lloyd Morgan's canon, against some of the more anthropomorphic work such as that of George Romanes, in which mental states had been freely attributed to animals. It was eagerly seized on by researchers such as Edward L. Thorndike (who had been studying cats' abilities to escape from puzzle boxes). However, most psychologists took up a position that is now called methodological behaviorism: they acknowledged that behavior was neither the only or the easiest method of observation in psychology, but held that it could be used to draw conclusions about mental states. Among well-known twentieth-century behaviorists taking this kind of position were Clark L. Hull, who described his position as neo-behaviorism, and Edward C. Tolman, who developed much of what would later become the cognitivist program. Tolman argued that rats constructed cognitive maps of the mazes they learned even in the absence of reward, and that the connection between stimulus and response (S->R) was mediated by a third term-the organism (S->O->R). His approach has been called, among other things, purposive behaviorism.

Methodological behaviorism remains the position of most experimental psychologists today, including the vast majority of those who work in cognitive psychology-so long as behavior is defined as including speech, at least non-introspective speech. With the rise of interest in animal cognition since the 1980s, and the more unorthodox views of Donald Griffin among others, mentalistic language including discussion of consciousness is increasingly used even in discussion of animal psychology, in both comparative psychology and ethology; however this is in no way inconsistent with the position of methodological behaviorism.

Skinner and Radical Behaviorism

Skinner, who carried out experimental work mainly in comparative psychology from the 1930s to the 1950s, but remained behaviorism's best known theorist and exponent virtually until his death in 1990, developed a distinct kind of behaviorist philosophy, which came to be called radical behaviorism. He is credited with having founded a new version of psychological science, which has come to be called behavior analysis or the experimental analysis of behavior after variations on the subtitle to his 1938 work *The Behavior of Organisms: An Experimental Analysis Of Behavior*.

Definition

Skinner was influential in defining radical behaviorism, a philosophy codifying the basis of his school of research (named the Experimental Analysis of Behavior, or EAB.) While EAB differs from other approaches to behavioral research on numerous methodological and theoretical points, radical behaviorism departs from methodological behaviorism most notably in accepting treatment of feelings, states of mind and introspection as existent and scientifically treatable. This is done by identifying them as something non-dualistic, and here Skinner takes a divide-and-conquer approach, with some instances being identified with bodily conditions or behavior, and others getting a more extended 'analysis' in terms of behavior. However, radical behaviorism stops short of identifying feelings as causes of behavior. Among other points of difference were a rejection of the reflex as a model of all behavior and a defense of a science of behavior complementary to but independent of physiology.

Experimental and Conceptual Innovations

This essentially philosophical position gained strength from the success of Skinner's early experimental work with rats and pigeons, summarised in his books The Behavior of Organisms (1938) and Schedules of Reinforcement (1957, with C. B. Ferster) and others. Of particular importance was his concept of the operant response, of which the canonical example was the rat's lever-press. In contrast with the idea of a physiological or reflex response, an operant is a class of structurally distinct but functionally equivalent responses. For example, while a rat might press a lever with its left paw or its right paw or its tail, all of these responses operate on the world in the same way and have a common consequence. Operants are often thought of as species of responses, where the individuals differ but the class coheres in its function--shared consequences with operants and reproductive success with species. This is a clear distinction between Skinner's theory and S-R theory.

Skinner's empirical work expanded on earlier research on trial-and-error learning by researchers such as Thorndike and Guthrie with both conceptual reformulations-Thorndike's notion of a stimulus-response 'association' or 'connection' was abandoned-and methodological ones-the use of the 'free operant', so called because the animal was now permitted to respond at its own rate rather than in a series of trials determined by the experimenter procedures. With this method, Skinner carried out substantial experimental work on the effects of different schedules and rates of reinforcement on the rates of operant responses made by rats and pigeons. He achieved remarkable success in training animals to perform unexpected responses, and to emit large numbers of responses, and to demonstrate many empirical regularities at the purely behavioral level. This lent some credibility to his conceptual analysis. It is largely his conceptual analysis that made his work much more rigorous than his peers, a point which can be seen clearly in his seminal work Are Theories of Learning Necessary? in which he criticizes what he viewed to be theoretical weaknesses then common in the study of psychology.

Relation to Language

As Skinner turned from experimental work to concentrate on the philosophical underpinnings of a science of behavior, his attention naturally turned to human language. His book Verbal Behavior (1957) laid out a vocabulary and theory for functional analysis of verbal behavior. The book was strongly criticized in a review by the linguist Noam Chomsky. Skinner did not himself respond in detail; he claimed that "[Chomsky] doesn't know what I am talking about and for some reason is unable to understand it".

What was important for a behaviorist analysis of human behavior was not language acquisition so much as the interaction between language and overt behavior. In an essay republished in his 1969 book Contingencies of Reinforcement, Skinner took the view that humans could construct linguistic stimuli that would then acquire control over their behavior in the same way that external stimuli could. The possibility of such "instructional control" over behavior meant that contingencies of reinforcement would not always produce the same effects on human behavior as they reliably do in other animals. The focus of a radical behaviorist analysis of human behavior therefore shifted to an attempt to understand the interaction between instructional control and contingency control, and also to understand the behavioral processes that determine what instructions are constructed and what control they acquire over behavior. Important figures in this effort have been Murray Sidman, A. Charles Catania, C. Fergus Lowe and Steven C. Hayes.

Molar versus Molecular Behaviorism

Skinner's view of behavior is most often characterized as a "molecular" view of behavior, that is each behavior can be decomposed into atomistic parts or molecules. This view is inaccurate when one considers his complete description of behavior as delineated in the 1981 article, Selection by Consequences and many other works. Skinner claims that a complete account of behavior involves an understanding of selection history at three levels: biology (the natural selection or phylogeny of the animal); behavior (the reinforcement history or ontogeny of the behavioral repertoire of the animal); and for some species, culture (the cultural practices of the social group to which the animal belongs).

This whole organism, with all those histories, then interacts with its environment. He often described even his own behavior as a product of his phylogenetic history, his reinforcement history (which includes the learning of cultural practices) interacting with the environment at the moment. Molar behaviorists, such as Howard Rachlin argue that behavior can not be understood by focusing on events in the moment. That is, they argue that a behavior can be understood best in terms of the ultimate cause of history and that molecular behaviorist are committing a fallacy by inventing a fictitious proximal cause for behavior. Molar behaviorists argue that standard molecular constructs such as "associative strength" are such fictitious proximal causes that simply take the place of molar variables such as rate of reinforcement. Thus, a molar behaviorist would define a behavior such as loving someone as exhibiting a pattern of loving behavior over time, there is no known proximal cause of loving behavior, only a history of behaviors (of which the current behavior might be an example of) that can be summarized as love. Recent experimental work shows quite clearly that behavior is affected both by molar variables (i.e., average rates of reinforcement) and molecular ones (e.g. time, preceding responses). What is needed is an understanding of the real-time dynamics of operant behavior, which will involve processes at both short and long time scales.

Behaviorism in Philosophy

Behaviorism is both a psychological movement and a philosophy of mind. The basic premise of radical behaviorism is that the study of behavior should be a natural science, such as chemistry or physics, without any reference to hypothetical inner states of organisms. Other varieties, such as theoretical behaviorism, permit internal states, but do not require them to be mental or have any relation to subjective experience. Behaviorism takes a functional view of behavior.

There are points of view within analytic philosophy that have called themselves, or have been called by others, behaviorist. In logical behaviorism (as held, e.g., by Rudolf Carnap and Carl Hempel), the meaning of psychological statements are their verification conditions, which consist of performed overt behavior. W. V. Quine made use of a type of behaviorism, influenced by some of Skinner's ideas, in his own work on language. Gilbert Ryle defended a distinct strain of philosophical behaviorism, sketched in his book The Concept of Mind. Ryle's central claim was that instances of dualism frequently represented 'category mistakes,' and hence that they were really misunderstandings of the use of ordinary language. Daniel Dennett likewise acknowledges himself to be a type of behaviorist.

It is sometimes argued that Ludwig Wittgenstein defended a behaviorist position, and there are important areas of overlap between his philosophy, logical behaviorism, and radical behaviorism (e.g., the beetle in a box argument). However, Wittgenstein was not a behaviorist, and his style of writing is sufficiently elliptical to admit of a range of interpretations. Mathematician Alan Turing is also sometimes considered a behaviorist, but he himself did not make this identification.

COGNITIVISM

In psychology, cognitivism is a theoretical approach in understanding the mind, which argues that mental function can be understood by quantitative, positivist and scientific methods, and that such functions can be described as information processing models.

Theoretical Approach

Cognitivism has two major components, one methodological, the other theoretical. Methodologically, cognitivism adopts a positivist approach and the belief that psychology can be (in principle) fully explained by the use of experiment, measurement and the scientific method. This is also largely a reductionist goal, with the belief that individual components of mental function (the 'cognitive architecture') can be identified and meaningfully understood. The second is the belief that cognition consists of discrete, internal mental states (representations or symbols) whose manipulation can be described in terms of rules or algorithms.

Cognitivism became the dominant force in psychology in the late-20th century, replacing behaviorism as the most popular paradigm for understanding mental function. Cognitive psychology is not a wholesale refutation of behaviorism, but rather an expansion that accepts that mental states exist. This was due to the increasing criticism towards the end of the 1950s of behaviorist models. One of the most notable criticisms was Chomsky's argument that language could not be acquired purely through conditioning, and must be at least partly explained by the existence of internal mental states.

Criticisms of Psychological Cognitivism

Cognitivism has been criticised in a number of ways.

Phenomenologists and hermeneutic philosophers have criticised the positivist approach of cognitivism for reducing individual meaning to what they perceive as measurements stripped of all significance. They argue that by representing experiences and mental functions as measurements, cognitivism is ignoring the context (cf contextualism) and, therefore, the meaning of these measurements. They believe that it is this personal meaning of experience gained from the phenomenon as it is experienced by a person (what Heidegger called being in the world) which is the fundamental aspect of our psychology that needs to be understood: therefore they argue that a context-free psychology is a contradiction in terms. They also argue in favour of holism: that positivist methods cannot be meaningfully used on something which is inherently irreducible to component parts. Hubert Dreyfus has been the most notable critic of cognitivism from this point of view. Humanistic psychology draws heavily on this philosophy, and practitioners have been among the most critical of cognitivism.

In the 1990s, various new theories emerged that challenged cognitivism and the idea that thought was best described as computation. Some of these new approaches, often influenced by phenomenological and post-modernist philosophy, include situated cognition, distributed cognition, dynamicism, embodied cognition. Some thinkers working in the field of artificial life (for example Rodney Brooks) have also produced non-cognitivist models of cognition.

The idea that mental functions can be described as information processing models has been criticised by philosopher John Searle and mathematician Roger Penrose who both argue that computation has some inherent shortcomings which cannot capture the fundamentals of mental processes.

- Penrose uses Gödel's incompleteness theorem (which states that there are mathematical truths which can never be proven in a sufficiently strong mathematical system; any sufficiently strong system of axioms will also be incomplete) and Turing's halting problem (which states that there are some things which are inherently non-computable) as evidence for his position.
- Searle has developed two arguments, the first (well known through his Chinese Room thought experiment) is the 'syntax is not semantics' argument-that a program is just syntax, understanding requires semantics, therefore programs (hence cognitivism) cannot explain understanding. It should be noted that such an argument presupposes the controversial notion of a private language. The second, which Searle now prefers but is

less well known, is his 'syntax is not physics' argument-nothing in the world is intrinsically a computer program except as applied, described or interpreted by an observer, so either everything can be described as a computer and trivially a brain can but then this does not explain any specific mental processes, or there is nothing intrinsic in a brain that makes it a computer (program). Detractors of this argument might point out that the same thing could be said about any concept-object relation, and that the brain-computer analogy can be a perfectly useful model if there is a strong isomorphism between the two. Both points, Searle claims, refute cognitivism.

Another argument against cognitivism is the problems of Ryle's Regress or the homunculus fallacy. Cognitivists have offered a number of arguments to refute these attacks.

The focused issues that interest cognitive psychologists consist of the inner mechanism of human thought and the processes of knowing. Cognitive psychologists have attempted to dig out the response to mental structures, such as what is saved and how it is recorded of course in our brain, once more to mental processes concerning how the integration and retrieval of information is operated. The theoretical assumptions in cognitive psychology lend instructional systems a hand in the design of efficient processing strategies for the learners to acquire knowledge, e.g. mnemonic devices to reduce the workload of the short-term memory, rehearsal strategies to maintain information, and the use of metaphors and analogies to relate meaning of the new information to prior knowledge.

COGNITIVE BEHAVIORAL THERAPY

A Cognitive Behavioral Therapy (CBT) is a psychotherapy based on modifying everyday thoughts and behaviors, with the aim of positively influencing emotions. The general approach developed out of behavior modification and Cognitive Therapy, and has become widely used to treat mental disorders. The particular therapeutic techniques vary according to the particular kind of client or issue, but commonly include keeping a diary of significant events and associated feelings, thoughts and behaviors; questioning and testing assumptions or habits of thoughts that might be unhelpful and unrealistic; gradually facing activities which may have been avoided; and trying out new ways of behaving and reacting. Relaxation and distraction techniques are also commonly included. CBT is widely accepted as an evidence-based, cost-effective psychotherapy for many disorders. It is sometimes used with groups of people as well as individuals, and the techniques are also commonly adapted for self-help manuals and, increasingly, for self-help software packages.

The Basics

Cognitive Behavioral Therapy (CBT) is based on the idea that how we think (cognition), how we feel (emotion) and how we act (behavior) all interact together. Specifically, our thoughts determine our feelings and our behavior. Therefore, negative- and unrealistic-thoughts can cause us distress and result in problems. One example could be someone who, after making a mistake, thinks "I'm useless and can't do anything right." This impacts negatively on mood, making the person feel depressed; the problem may be worsened if the individual reacts by avoiding activities. As a result, a successful experience becomes more unlikely, which reinforces the original thought of being "useless." In therapy, the latter example could be identified as a self-fulfilling prophecy or "problem cycle," and the efforts of the therapist and client would be directed at working together to change this. This is done by addressing the way the client thinks in response to similar situations and by developing more flexible thought patterns, along with reducing the avoidance of activities. If, as a result, the client escapes the negative thought pattern, the feelings of depression may be relieved. The client may then become more active, succeed more often, and further reduce feelings of depression.

Thoughts as the Cause of Emotions

With thoughts stipulated as being the cause of emotions rather than the outcome or by-product, cognitive therapists reverse the causal order more generally used by psychotherapists. Therefore, the therapy is to identify those irrational or maladaptive thoughts that lead to negative emotion and identify what it is about them that is irrational or just not helpful; this is done in an effort to reject the distorted thoughts and replace them with more realistic alternative thoughts.

Cognitive Behavioral Therapy is not an overnight process. Even after patients have learned to recognize when and where their thought processes go awry, it can take months of concerted effort to replace an irrational thought process or habit with a more reasonable, salutary one.

The cognitive model says that a person's core beliefs (often formed in childhood) contribute to 'automatic thoughts' that pop up in every day life in response to situations. Cognitive Therapy practitioners hold that clinical depression is typically associated with negatively biased thinking and irrational thoughts. Cognitive Behavioral Therapy is often used in conjunction with mood stabilizing medications to treat bipolar disorder. Its application in treating schizophrenia along with medication and family therapy is recognized by the NICE guidelines within the British NHS.

Cognitive Behavioral Therapy (CBT)

Cognitive Behavioral Therapy (CBT) can be seen as an umbrella term for many different therapies that share some common elements. While similar views of emotion have existed for millennia, the earliest form of Cognitive Behavior Therapy was developed by Albert Ellis in the early 1950s. Ellis eventually called his approach Rational Emotive Behavioral Therapy, or REBT, as a reaction against popular psychoanalytic and increasingly humanistic methods at the time. Aaron T. Beck independently developed another CBT approach, called Cognitive Therapy, in the 1960s. Cognitive Therapy rapidly became a favorite intervention to study in psychotherapy research in academic settings. In initial studies, it was often contrasted with behavioral treatments to see which was most effective. However, in recent years, cognitive and behavioral techniques have often been combined into cognitive behavioral treatment. This is arguably the primary type of psychological treatment being studied in research today.

Cognitive Behavioral Group Therapy (CBGT) is a similar approach in treating mental illnesses, based on the protocol by

Richard Heimberg. In this case, clients participate in a group and recognize they are not alone in suffering from their problems.

A sub-field of Cognitive Behavior Therapy used to treat Obsessive Compulsive Disorder makes use of classical conditioning through extinction (a type of conditioning) and habituation. (The specific technique, Exposure with Response Prevention (ERP) has been demonstrated to be more effective than the use of medication--typically SSRIs--alone.) CBT has also been successfully applied to the treatment of Generalized Anxiety Disorder, health anxiety, Social phobia and Panic Disorder. In recent years, CBT has been used to treat symptoms of schizophrenia, such as delusions and hallucinations. This use has been developed in the UK by Douglas Turkington and David Kingdon.

Other types of Cognitive Behavioral Therapy include Dialectical Behavior Therapy, Self-Instructional Training, Schema-Focused Therapy and many others.

CBT has a good evidence base in terms of its effectiveness in reducing symptoms and preventing relapse. It has been clinically demonstrated in over 400 studies to be effective for many psychiatric disorders and medical problems for both children and adolescents. It has been recommended in the UK by the National Institute for Health and Clinical Excellence as a treatment of choice for a number of mental health difficulties, including post-traumatic stress disorder, OCD, bulimia nervosa and clinical depression. Cognitive Behavioral Therapy most closely allies with the Scientist-Practitioner Model of Clinical Psychology, in which clinical practice and research is informed by a scientific perspective; clear operationalization of the "problem" or "issue;" an emphasis on measurement (and measurable changes in cognition and behavior); and measurable goal-attainment.

Depression

Negative thinking in depression can result from biological sources (i.e., endogenous depression), modelling from parents, peers or other sources. The depressed person experiences negative thoughts as being beyond their control: the negative thought pattern can become automatic and self-perpetuating.

Negative thinking can be categorized into a number of common patterns called "cognitive distortions." The cognitive therapist provides techniques to give the client a greater degree of control over negative thinking by correcting these distortions-or correcting thinking errors that abet the distortions, in a process called cognitive restructuring. Negative thoughts in depression are generally about one or more of three areas: negative view of self, negative view of the world and negative view of the future. These constitute what Beck called the "cognitive triad."

Attributional Style

An approach to depression based upon attribution theory in social psychology is related to the concept of attributional style. First put forth by Lyn Abramson and her colleagues in 1978, this approach argues that depressives have a typical attributional style-they tend to attribute negative events in their lives to stable and global characteristics of themselves. There is considerable evidence that depressives do exhibit such an attributional style, but it is important to remember that Abramson et al. do not claim that an attributional style of this nature is necessarily going to cause depression-only that it will lead to clinical depression if combined with a negative event. This theory is sometimes known as a revised version of learned helplessness theory. In 1989, this theory was challenged by Hopelessness Theory. This theory emphasised attributions to global and stable factors, rather than, as in the original model, internal attributions. Hopelessness Theory also emphasises that beliefs about the consequences of events and rated importance of events may be at least as important in understanding why some people react to negative events with clinical depression as are causal attributions.

Irrational Beliefs

A major aid in cognitive therapy is what Albert Ellis called the ABC Technique of Irrational Beliefs. The first three steps analyse the process by which a person has developed irrational beliefs and may be recorded in a three-column table.

- *A-Activating Event or Objective Situation:* The first column records the objective situation, that is, an event that ultimately leads to some type of high emotional response or negative dysfunctional thinking.

- *B-Beliefs:* In the second column, the client writes down the negative thoughts that occurred to them.
- *C-Consequence:* The third column is for the negative feelings and dysfunctional behaviors that ensued. The negative thoughts of the second column are seen as a connecting bridge between the situation and the distressing feelings. The third column C is next explained by describing emotions or negative thoughts that the client thinks are caused by A. This could be anger, sorrow, anxiety, etc.

  For example, Gina is upset because she got a low mark on a math test. The Activating event, A, is that she failed her test. The Belief, B, is that she must have good grades or she is worthless. The Consequence, C, is that Gina feels depressed.
- *Reframing:* After irrational beliefs have been identified, the therapist will often work with the client in challenging the negative thoughts on the basis of evidence from the client's experience by reframing it, meaning to re-interpret it in a more realistic light. This helps the client to develop more rational beliefs and healthy coping strategies.

From the example above, a therapist would help Gina realize that there is no evidence that she must have good grades to be worthwhile, or that getting bad grades is awful. She desires good grades, and it would be good to have them, but it hardly makes her worthless. If she realizes that getting bad grades is disappointing, but not awful, and that it means she is currently bad at math or at studying, but not as a person, she will feel sad or frustrated, but not depressed. The sadness and frustration are likely healthy negative emotions and may lead her to study harder from then on.

Effectiveness of CBT with or without Drugs for Depression

A large-scale study in 2000 showed substantially higher results of response and remission when a form of cognitive behavior therapy and an anti-depressant drug were combined than when either method was used alone.

The effectiveness of combination therapy is endorsed by the Australian depressioNet group: Currently the most effective treatment for major (clinical) depression is considered to be a combination of antidepressant medication and Cognitive Behavioral Therapy.

For more general results confirming that CBT alone can provide lower but nonetheless valuable levels of relief from depression, and result in increased ability for the patient to stay in employment.

The typical short-term success rate for CBT is about 50%. In other words, if 100 people attend up to sixteen weekly sessions one-on-one lasting one hour each, some will drop out but within four months 50 people will have lost their psychiatric symptoms over and above those who would have done so anyway. After recovery, people who suffered from anxiety are unlikely to relapse.... So how much depression can a course of CBT relieve, and how much more work will result? One course of CBT is likely to produce 12 extra months free of depression. This means nearly two months more of work.

The American Psychiatric Association Practice Guidelines (April 2000) indicated that among psychotherapeutic approaches, cognitive behavioral therapy and interpersonal therapy had the best-documented efficacy for treatment of major depressive disorder, although they noted that rigorous evaluative studies had not been published.

CBT with Children and Adolescents

The use of CBT has been extended to children and adolescents with good results. It is often used to treat depression, anxiety disorders, and symptoms related to trauma and Post Traumatic Stress Disorder. Significant work has been done in this area by Mark Reinecke and his colleagues at Northwestern University in the Clinical Psychology program in Chicago.

CBT has been used with children and adolescents to treat a variety of conditions with good success.

CBT is also used as a treatment modality for children who have experienced Complex Post Traumatic Stress Disorder, chronic maltreatment, and Post Traumatic Stress Disorder. It would be one component of treatment for children with C-PTSD, along with a variety of other components, which are discussed in the Complex Post Traumatic Stress Disorder article. In addition, many approaches to treating such children, such as Dyadic Developmental Psychotherapy incorporate Cognitive therapy methods and principles into treatment.

Computerised CBT

As the name suggests, this is a computerised form of CBT, in which the user interacts with computer software (either on a PC, or sometimes via a voice-activated phone service), instead of face-to-face with a therapist.

Computerised CBT is not a replacement for face-to-face therapy but can provide an option for patients, especially in light of the fact that there are not always therapists available, or the cost can be prohibitive. Computerised CBT is clinically proven and drug-free. For people who are feeling depressed and withdrawn, the prospect having to speak to someone about their deepest problems can be off-putting. In this respect, CCBT (especially if delivered online) can be a good option.

It has been proven to be effective in Randomised Controlled Trials, and in February 2006 the National Institute of Health and Clinical Excellence (NICE) recommended that CCBT should be made available for use within the NHS across England and Wales, for patients presenting with mild/moderate depression, rather than immediately opting for medication (i.e. anti-depressant pills). That guidance can be read here

A new Government initiative for tackling Mental Health issues, Choices In Mental Health has recently been launched by the Care Services Improvement Partnership. This confirms Primary Care Trust(PCT) responsibilities in delivering the NICE Technology Appraisal on CCBT. National Director for Mental Health, Professor Louis Appleby CBE has confirmed that by 31st March 2007 PCTs should have CCBT products in place and the NICE Guidelines should be met.

8

Therapies in Psychology

EXISTENTIAL THERAPY

Existential psychotherapy is partly based on the existential belief that human beings are alone in the world. This aloneness leads to feelings of meaninglessness which can be overcome only by creating one's own values and meanings. We have the power to create because we have the freedom to choose. In making our own choices we assume full responsibility for the results and blame no one but ourselves if the result is less than what was desired.

The psychotherapist helps his or her patients/clients along this path: to discover why the patient/client is overburdened by the anxieties of aloneness and meaninglessness, to find new and better ways to manage these anxieties, to make new and healthy choices, and to emerge from therapy as a free and sound human being.

Existential therapy focuses on the development of a patient/client's self-awareness by looking deeply into the issues of our aloneness, meaninglessness, and mortality. The therapist emphasizes the patient/client's ability to freely make choices in the present, not under the influence of deterministic aspects or past conditioning.

The existentialist attempts to convert meaninglessness into meaningfulness, giving the patient/client the courage to make his or her own healthy choices and to lead a socially rewarding life. Existential therapists have their own unique views about human nature, mental dysfunction, wellness, and therapeutic techniques.

View of the Human Mind

Although humans are essentially alone in the world, we long to be connected to others. We want to have meaning in their lives while they have meaning in ours, but ultimately we must come to realize that we cannot depend on others for our validation, and with that realization we finally acknowledge and understand that we are fundamentally alone. The result of this revelation is anxiety in the knowledge that our validation must come from within and not from others. This anxiety is a positive thing, since it makes us aware of the limits of human existence and this makes us capable of reflecting on life instead of living thoughtlessly.

Because we are alienated and isolated our lives are also meaningless. Nothing exists which is greater than ourselves, therefore there are no external sources of values and absolutes from which we can draw. Taken to an extreme we might conclude that there is nothing for which to live, however, there is hope in the possibility of creating our own values and our own meanings and applying them to our condition, giving us feelings of significance and purpose that are strong enough to carry us through life. This freedom we have to choose our own values is another source of anxiety: we must summon the requisite strength and courage to choose our life-meaning and hold fast to it, undoubtedly a task which many find difficult.

Human beings are also mortal. As we come to grips with the fact that our lives are limited, we develop even more anxiety: we are afraid of death. The knowledge that at some point in the future we will cease to be, while frightening, is at the same time invigorating because it is relevant right now. The juxtaposition of life and death is one thing that does give us some certainty.

Finally, humans are responsible. Being isolated, alone, and free to choose means that one cannot assign blame for his or her problems to someone else. The individual alone makes the choices and therefore is responsible for the outcomes. At any point we are free to make different choices and thus re-invent ourselves; we are at once the architect, the planner, and the builder of our lives, throughout our lives.

Psychological Dysfunction

There is no such thing as psychological dysfunction or being ill in the existential view. Every way of being is merely an expression of how one chooses to live one's life. However one may feel unable to come to terms with the anxiety of being alone in the world. If so an existential psychotherapist can assist one in accepting these feelings rather than trying to change them as if there is something wrong. Everyone has the freedom to choose how they are going to be in life, however this may go unexercised because making changes is difficult; it may appear easier and safer not to make decisions that you will be responsible for. Many people will remain unaware of alternative choices in life for various societal reasons.

The Good Life

It is possible for people to face the anxieties of life head-on and embrace the human condition of aloneness, to revel in the freedom to choose and take full responsibility for their choices. They courageously take the helm of their lives and steer in whatever direction they choose; they have the courage to be. One does not need to arrest feelings of meaninglessness, but can choose new meanings for their lives. By building, by loving, and by creating one is able to live life as one's own adventure. One can accept one's own mortality and overcome fear of death.

EXISTENTIAL THERAPY

The existentially-oriented psychotherapist guides his or her patients/clients to confront life's anxieties. If the patient/client has not been fully exercising the freedom to choose, the counselor will lead a discovering into how and why he or she is stuck. Perhaps the patient/client has been allowing others to make the important decisions which he or she alone should be making. Possibly the patient/client is afraid to take the risks required to grow and is instead choosing an easy and non-threatening path. The counselor will encourage his or her patient/clients to reflect on the aloneness and meaninglessness of life, and to understand that they must find their own ways to cope with these anxieties. The counselor does not try to eliminate these anxieties, but instead encourages the patient/client to face them head-on. Alternative paths can be explored together. The risks entailed with these paths can be evaluated, and then the patient/client will be-able to make new, more authentic choices. The existential counselor is not overly concerned with the patient's/client's past. Instead, the emphasis is on the choices to be made in the present. The counselor and the client may reflect upon how the client has answered life's questions in the past, but then attention shifts to searching for a new and increased awareness in the present and enabling a new freedom and responsibility to act.

FAMILY THERAPY

Family therapy, also referred to as couple and family therapy and family systems therapy, and earlier generally referred to as marriage therapy, is a branch of psychotherapy that works with families and couples in intimate relationships to nurture change and development. It tends to view these in terms of the systems of interaction between family members. It emphasizes family relationships as an important factor in psychological health. As such, family problems have been seen to arise as an emergent property of systemic interactions, rather than to be blamed on individual members.

Family therapists may focus more on how patterns of interaction maintain the problem rather than trying to identify the cause, as this can be experienced as blaming by some families. It assumes that the family as a whole is larger than the sum of its parts.

Most practitioners are "eclectic", using techniques from several areas, depending upon the client(s). Family therapy practitioners come from a range of professional backgrounds, and some are specifically qualified or licensed/registered in family therapy (licensing is not required in some jurisdictions and requirements vary from place to place). In the UK, family therapists are usually psychologists, nurses, psychotherapists, social workers, or counselors who have done further training in family therapy, either a diploma or an M.Sc..

Multicultural and inter-cultural approaches are being developed.

Family therapy has been used effectively where families, and or individuals in those families experience or suffer:

- serious psychological disorders (e.g. schizophrenia, addictions and eating disorders);
- interactional and transitional crises in a family's life cycle (e.g. divorce);
- as a support of other psychotherapies and medication.

Methodology

It uses a range of counseling and other techniques including:

- psychotherapy
- systems theory
- communication theory
- systemic coaching

The number of sessions depends on the situation, but the average is 5-20 sessions. The basic theory of family therapy is derived mainly from object relations theory, cognitive psychotherapy, systems theory and narrative approaches.

A family therapist usually meets several members of the family at the same time ("conjoint family therapy" is used in the approach of Virginia Satir.) This has the advantage of making differences between the ways family members perceive mutual relations as well as interaction patterns in the session apparent both for the therapist and the family. These patterns frequently mirror habitual interaction patterns at home, even though the therapist is now incorporated into the family system. Therapy interventions usually focus on relationship patterns rather than on analyzing impulses of the unconscious mind or early childhood trauma of individuals as a Freudian therapist would do.

Family therapy is really a way of thinking, an epistemology rather than about how many people sit in the room with the therapist. Family therapists are relational therapists; they are interested in what goes between people rather than in people.

Depending on circumstances, a therapist may point out to the family interaction patterns that the family might have not noticed; or suggest different ways of responding to other family members. These changes in the way of responding may then trigger repercussions in the whole system, leading to a more satisfactory system state.

FEMINIST THERAPY

Feminist Therapy is a set of related therapies arising from the disparity between the origin of most psychological theories (which have male authors) and the majority of people seeking counseling being female. It focuses on societal, cultural, and political causes and solutions to issues faced in the counseling process. It openly encourages the client to participate in the world in a more social and political way.

Shared Concepts in Feminist Therapy

Feminist therapies arose from various contexts, authors, and historical perspectives. However, all of them can be considered to share some basic characteristics.

- The person is political
- Personal and social identities are interdependent
- Conventional distress and "mental illness" concepts are challenged
- The role of oppression is central
- The counseling relationship is egalitarian
- Women's perspectives are valued

Principal Styles of Feminist Therapies

The principal styles noted here correspond to different types of feminism.

- Liberal
- Cultural
- Radical
- Socialist
- Postmodern
- Women of color

- Lesbian
- Global-international

Goals of Therapy

Five principal goals of therapy exist:

- Equality
- Balancing independence and interdependence
- Empowerment
- Self-nurturance
- Valuing diversity

Therapeutic Strategies

The following strategies are typical of feminist therapy.

- Empowerment
- Self-disclosure
- Gender-role analysis
- Gender-role intervention
- Power analysis and power intervention
- Bibliotherapy
- Assertiveness training
- Reframing and relabeling
- Group work
- Social action

GESTALT THERAPY

Gestalt Therapy is a psychotherapy which focuses on here-and-now experience and personal responsibility. It was co-founded by Fritz Perls, Laura Perls and Paul Goodman in the 1940s-1950s.

General Description

The objective of Gestalt Therapy, in addition to helping the client overcome symptoms, is to enable the her-him to become more fully and creatively alive and to be free from the blocks and unfinished issues which may diminish optimum satisfaction, fulfillment, and growth. Thus, it falls in the category of humanistic psychotherapies.

Gestalt therapy (GT) has its roots in psychoanalysis. It was part of a continuum moving from the early work of Freud, to the later Freudian ego analysis, to Wilhelm Reich and his notion of character armor. To this was added the insights of academic gestalt psychology about perception, gestalt formation and the tendency of organisms to complete the incomplete gestalt, to form "wholes" in experience.

There were additional influences from existentialism, particularly the I-thou relationship as it applies to therapy, and the notion of personal choice and responsibility.

Central to Fritz and Laura Perls' modifications of psychoanalysis was the concept of "dental or oral aggression". Dental aggression is discussed in "Ego, Hunger, and Aggression" (1944), Fritz Perl's first book, to which Laura Perls contributed, in her own words: "at least two chapters which I wrote completely: the chapters on the dummy complex and the one on insomnia." Laura Perls was credited in the preface to the first edition, but in subsequent editions her name was left out. In "Ego, Hunger and Aggression", the Perls suggested that when the infant develops teeth, he/she has the capacity to chew, to break apart food, and by analogy experience, to taste, accept, reject, assimilate. This was opposed to Freud's notion that only introjection takes place in early experience. Thus the Perls made "assimilation", as opposed to "introjection", a focal theme in their work, and the prime means by which growth occurs in therapy.

In contrast to the psychoanalytic stance in which the "patient" introjects the (presumably more healthy) attitudes/ interpretations of the analyst, in Gestalt Therapy the client must "taste" his/her experience, and either accept or reject, but not introject, or "swallow whole". Hence, the emphasis is on avoiding interpretation and encouraging discovery. This is the key point in the divergance of GT from traditional psychoanalysis--growth occurs through gradual assimilation of experience in a natural way, rather than by accepting the interpretations of the analyst; thus, the therapist should not interpret, but lead the client to discover for him or herself.

The Gestalt therapist contrives experiments that lead the client to greater awareness and fuller experience of his/her possibilities. Experiments can be focussed on undoing projections or retroflections. They can work to help the client with closure of unfinished gestalts ("unfinished business" such as unexpressed emotions towards somebody in the client's life). There are many kinds of experiments that might be therapeutic. But the essence of the work is that it is experiential rather than interpretive, and in this way distinguishes itself from the psychoanalytic. Perls also derived much from Reich's emphasis on how defenses are embodied, and therefore paid a great deal of attention to nonverbal behavior. This was consonant with Laura Perls' background in dance and movement therapy.

A core concept in Gestalt therapy is the unifying idea of "contact". Contact is where one person meets another person, or meets the outside world. Thus, there can be physical contact, but mostly what is meant by the term is metaphoric. If contact is not interfered with by what Perls-Goodman called disturbances of the contact boundary, the individual can grow, through assimilation of new experiences.

In therapy, the client is encouraged to experience his or her own feelings and behaviours in the here and now, and attention is brought to bear on the way contact is interrupted. The way in which he or she interrupts contact with the present environment is considered to be a significant factor in creating and maintaining dysfunctional patterns of behavior.

Some of the contact interruptions occur through projection (seeing outside one's self what belongs to one's self), introjection (swallowing whole instead of assimilating, chewing, digesting); retroflection (directing impulses towards the self that rightly should be directed to the other, as in anger directed toward self causing depression or psychosomatic symptoms); confluence (dissolving the self-other boundary and merging with the other).

A disturbance introduced by Miriam and Erv Polster is "deflection," referring to a means of avoiding contact by jumping around from one thing to another and never staying in the same place for very long. All of these disturbances have a pathological and a non-pathological aspect. It is appropriate for the infant and mother to become confluent, for example, or two lovers, but inappropriate for client and therapist. When the latter pair becomes confluent, there can be no growth because there is no boundary at which the one can contact the other; the client will not be able to learn anything new because the therapist is simply an extension of the client, so to speak.

Another important aspect of contact, one which can be found in Perls, Hefferline and Goodman, but was neglected by almost everyone except the unfortunately writing-phobic Isadore From, was that contact-making can be deconstructed into stages. One can look at a pathology of the contact making function and relate this to personality disturbances. An example would be that some people grasp onto people and experiences too quickly, without orienting themselves to what is out there. This would be a disturbance at the "fore-contact" stage, and would be part of an impulsive personality style. This aspect of GT is a neglected part of the theory that connect GT to the original work of the Gestalt psychologists on phenomenology and perception.

By focusing the individual on how contact-making occurs or is disturbed, new insights can be made and the fluid process of adequate contacting resumed.

Background

Fritz Perls was a German psychoanalyst who fled with his wife Lore to South Africa to escape Nazi oppression. After the war the couple emigrated to New York City, which had become by the late 1940s and early 1950s, a center of intellectual, artistic, and political experimentation.

The seminal work was Gestalt Therapy: Excitement and Growth in the Human Personality, published in 1951; co-authored by Fritz Perls, Paul Goodman, and Ralph Hefferline (a university psychology professor, and sometime patient of Fritz Perls). As it turns out, most of the original part II of the book was written by Paul Goodman from the notes of Fritz Perls, and contains the meat of the theory. It was supposed to go first. The publishers decided that Part II, written by

Hefferline, fit more into the nascent self-help ethos of the day, and made it Part I, making for a less interesting introduction to the theory. Isadore From, a leading early theorist of Gestalt Therapy, taught Part II for an entire year to his students, going through it phrase by phrase.

Gestalt therapy had a variety of psychological and philosophical influences, and in addition was a response to the social forces of its day. It is a therapeutic approach which is holistic (mind/body/culture) present-centered, and related to existential therapy in its emphasis on personal responsibility for action, and on the valuing of the I-thou relationship in therapy. (In fact, its creators considered calling Gestalt Therapy existential-phenomenological therapy.) "The I and thou in the Here and Now," was one Gestalt therapist's semi-humorous mantra.

Fritz and Lore (now Laura) founded the first Gestalt Institute in New York City in 1952. Isadore From became a patient, first of Fritz and then of Laura. Pretty quickly Fritz anointed Isadore a trainer and also gave him some patients. Isadore lived in New York until his death at 75 in 1993 and was known world-wide for his philosophical and intellectually rigorous take on Gestalt Therapy. A brilliant, witty and sometimes caustic man, From was definitely the philosopher of the first-generation Gestalt therapists. He was also acknowledged to be a supremely gifted clinician. Unfortunately, he was phobic of writing. The only things committed to paper are a few interviews, but they are worth reading.

Jim Simkin was a psychologist who also became a client of Perls and then a co-trainer with Perls in California. Simkin was responsible for Perls coming to California where he attempted to begin a psychotherapy practice. Ultimately, being a peripatetic trainer and workshop leader was a better fit for Fritz' personality. Jim and Fritz co-led some of the early (for California) training groups at Esalen.

The late 1950s-1960s movement toward personal growth and the human potential movement fed into and was itself influenced by Gestalt Therapy. With all these disparate influences, Gestalt therapy somehow became a "coherent gestalt", which is the Gestalt psychologists' term for a perceptual unit that holds together and forms a unified form. In the 1960s Perls became infamous for his public workshops at Esalen Institute in Big Sur. Isadore From referred to some of Fritz' several day workshops as "hit-and-run" therapy because of its emphasis on showmanship with little or no follow-through, but Fritz never considered these workshops to be true therapy. Jim Simkin went from co-leading training groups with Fritz to purchasing a property next to Esalen and starting his own training center, which he ran until his death in 1984. Here he refined his precise laser-like version of Gestalt Therapy, training psychologists, psychiatrists, and LCSW's within a very rigorous residential training model.

When Fritz Perls left New York City for California, there began to be a split between those who saw Gestalt Therapy as a therapeutic approach with great potential (this view was best represented by Isadore From, who practiced and taught mainly in New York, and by the members of the Cleveland Institute, co-founded by From) and those who saw Gestalt Therapy not just as a therapeutic modality but as a way of life. The East Coast, New York-Cleveland axis was often appalled by the notion of Gestalt Therapy leaving the consulting room and becoming a way-of-life in the West Coast of the 1960s.

The split continues between what has been called "East Coast" GT and "West Coast" GT. However, the way-of-life view seems to be fading as people move on from the 1960s. Esalen is still functioning in Big Sur. The widow of Esalen's co-founder Dick Price, Christine Price, continues to hold Gestalt workshops there. In 1969 Fritz Perls left the USA to start a Gestalt community at Lake Cowichan on Vancouver Island, Canada. He died almost a year later on 14 March 1970 in Chicago. One member of the Gestalt community was Barry Stevens. Her book about that phase of her life became very popular ("Don' t Push the River."). She developed her own form of Gestalt therapy body work which is essentially a concentration on the awareness of body processes.

Other Gestalt Therapists

Although Gestalt Therapy reached its zenith in the late 1970s and early 1980s and has since waned in popularity, its contributions have become assimilated into current schools of therapy, sometimes in unlikely places. For example, Acceptance and Commitment Therapy (ACT) shares much from Gestalt Therapy yet is considered to be a cognitive behavioral approach. Also, mindfulness is a buzzword as of 2006, yet much of mindfulness work is connected to Gestalt Therapy's emphasis on the flow of experience and awareness.

Dan Rosenblatt led Gestalt training groups in Japan for 7 years. Rosenblatt (b. 1925) was part of the early group around Laura. A Harvard-trained psychologist and intellectual, he practiced Gestalt therapy for over 35 years in Manhattan, seeing 30 patients a week in individual therapy and doing groups almost every evening. He did training workshops in Germany, the Netherlands, Japan, New Zealand, Italy for many years. Rosenblatt, who also wrote several books on Gestalt therapy, exemplifies the Gestalt therapist as practicing clinician, rather than would-be guru.

All of these therapists had their own distinctive styles, but always with Gestalt Therapy's focus on immediate experience as a central theme. And unlike Fritz Perls, who Isadore From persisted in calling Frederick Perls, these first generation Gestalt therapists maintained thriving therapy practices, mostly in one location, for many years and Stewart Kiritz followed with public workshops and training workshops from 1997 through 2005.

Gestalt Therapy is a very useful process for therapists-in-training of any persuasion because of its focus on the person of the therapist, barriers to full contact with others, self-awareness. And graduate students still seem to seek it out, even though it is not as recognized by the establishment as it once was.

Principal influences

- Otto Rank's invention of "here-and-now" therapy and Rank's post-Freudian book Art and Artist (1932), both of which strongly influenced Paul Goodman.
- Wilhelm Reich's psychoanalytic developments, especially the concept of character armor and its focus on the body.

- Jacob Moreno's Psychodrama, principally the development of enactment techniques for the resolution of psychological conflicts
- Max Wertheimer's Gestalt Psychology, which this therapy derives its name from, influences the application of the concepts about perception to a broader theory about human needs and the relation of persons with their surroundings
- Kurt Goldstein's holistic theory of the organism, based on Gestalt theory.
- Martin Buber's philosophy of relationship and dialogue ("I-Thou")
- Kurt Lewin's field theory as applied to the social sciences and group dynamics
- European phenomenology of Franz Brentano, Edmund Husserl, Martin Heidegger, and Maurice Merleau-Ponty
- The existentialism of Kierkegaard over that of Sartre, rejecting nihilism
- Carl Jung's psychology, particularly the polaritics concept
- Some elements from Zen Buddhism
- American pragmatism of William James, George Herbert Mead, and John Dewey

9

Humanistic Psychology

Humanistic psychology is a school of psychology that emerged in the 1950s in reaction to both behaviorism and psychoanalysis. It is explicitly concerned with the human dimension of psychology and the human context for the development of psychological theory. These matters are often summarized by the five postulates of Humanistic Psychology given by James Bugental (1964), mainly that:

1. Human beings cannot be reduced to components.
2. Human beings have in them a uniquely human context.
3. Human consciousness includes an awareness of oneself in the context of other people.
4. Human beings have choices and responsibilities.
5. Human beings are intentional, they seek meaning, value and creativity.

THE DEVELOPMENT

The humanistic approach has its roots in existentialist thought. It is also sometimes understood within the concept of the three different forces of psychology; behaviorism, psychoanalysis and humanism. Behaviorism grew out of Ivan Pavlov's work with the conditioned reflex, and laid the foundations for academic psychology in the United States associated with the names of John B. Watson and B.F. Skinner.

This school was later called the science of behavior. Abraham Maslow later gave behaviorism the name "the first force". The "second force" came out of Freud's research of psychoanalysis, and the psychologies of Alfred Adler, Erik Erikson, Carl Jung, Erich Fromm, Karen Horney, Otto Rank, Melanie Klein, Harry Stack Sullivan, and others. These theorists focused on the depth of the human psyche, which they stressed, must be combined with those of the conscious mind in order to produce a healthy human personality.

In the late 1950's, psychologists concerned with advancing a more holistic vision of psychology convened two meetings in Detroit, Michigan. These psychologists, including Abraham Maslow, Carl Rogers, and Clark Moustakas, were interested in founding a professional association dedicated to a psychology that focused on uniquely human issues, such as the self, self-actualization, health, hope, love, creativity, nature, being, becoming, individuality, and meaning-in short, the understanding of what it means to be human.

These preliminary meetings eventually led to other developments, which culminated in the description of humanistic psychology as a recognizable "third force" in psychology (along with behaviorism and psychoanalysis). Significant developments included the launch of the Journal of Humanistic Psychology in 1961 and the formation of the Association for Humanistic Psychology (AHP) in 1963. Subsequently, graduate programs in humanistic psychology at institutions of higher learning grew in number and enrollment. In 1971, humanistic psychology as a field was recognized by the American Psychological Association (APA) and granted its own division (Division 32) within the APA. Division 32 publishes its own academic journal called The Humanistic Psychologist (Aanstoos, Serlin & Greening, 2000).

The major theorists considered to have prepared the ground for Humanistic Psychology are Abraham Maslow, Carl Rogers and Rollo May. The work of Wilhelm Reich, who postulated an essentially 'good', healthy core self, in contrast to Freud, was an early influence, specially his "Character Analysis" (1933). Other noteworthy inspirers and leaders of the movement include Roberto Assagioli, Gordon Allport, Medard Boss, Martin Buber, R. D. Laing, Fritz Perls, Anthony Sutich, Erich Fromm, Kurt Goldstein, Clark Moustakas, Lewis Mumford and James Bugental (Aanstoos, Serlin & Greening, 2000).

Epistemology

Humanistic psychology usually prefers qualitative research methods over other epistemological approaches (Clay, 2002). This is part of the field's "human science" approach to psychology; an emphasis on the actual lived experience of persons (Aanstoos, Serlin & Greening, 2000). Scientifically, the humanistic field views the usage of quantitative methods in the study of the human mind and behaviour as misguided. This is in direct contrast to cognitivism (which aims to apply the scientific method to the study of psychology), an approach of which humanistic psychology has been strongly critical. Instead, the discipline stresses a phenomenological view of human experience, seeking to understand human beings and their behavior by conducting qualitative research.

Counseling and Therapy

Humanistic psychology includes several approaches to counseling and therapy. Among the earliest approaches we find the developmental theory of Abraham Maslow, emphazising a hierarchy of needs and motivations; the existential psychology of Rollo May acknowledging human choice and the tragic aspects of human existence; and the person-centered or client-centered therapy of Carl Rogers, which is centered around the clients' capacity for self-direction and understanding of his/her own development (Clay, 2002).

Other approaches to humanistic counselling and therapy include Gestalt therapy, humanistic psychotherapy, depth therapy, holistic health, encounter groups, sensitivity training, marital and family therapies, body work, and the existential psychotherapy of Medard Boss (Aanstoos, Serlin & Greening (2000).

Self-help is also included in humanistic psychology. Ernst & Goodison (1981) describe using some of the main humanistic approaches in self-help groups. Co-counselling, which is a purely self-help approach, is regarded as coming within humanistic psychology. Humanistic theory has had a strong influence on other forms of popular therapy, including Harvey Jackins' Re-evaluation Counselling and the work of Carl Rogers.

As mentioned by Clay (2002) Humanistic psychology tends to look beyond the medical model of psychology in order to open up a nonpathologizing view of the person. This usually implies that the therapist downplays the pathological aspects of a person's life in favour of the healthy aspects. A key ingredient in this approach is the meeting between therapist and client and the possibilities for dialogue. The aim of much humanistic therapy is to help the client approach a stronger and more healthy sense of self, also called self-actualization (Aanstoos, Serlin & Greening, 2000; Clay, 2002). All this is part of Humanistic psychology's motivation to be a science of human experience, focusing on the actual lived experience of persons (Aanstoos, Serlin & Greening (2000).

Criticism and Debate

Criticism of the field has come from Isaac Prilleltensky (1992) who argues that humanistic psychology-inadvertently-is affirming the social and political status quo, and therefore has remained fairly silent about social change.

Further, in their review of different approaches to positive psychology, Seligman & Csikszentmihalyi (2000) note that the early incarnations of humanistic psychology lacked a cumulative empirical base, and that some directions encouraged self-centeredness. However, according to mainstream humanistic thinkers, humanistic psychology must not be understood to promote such ideas as the narcissism, egotism, or selfishness (Bohart & Greening, 2001). The association of humanistic discourse with narcissistic and overly optimistic worldviews is a misreading of humanistic theory.

In their response to Seligman & Csikszentmihalyi (2000), Bohart & Greening (2001) note that along with pieces on self-actualization and individual fulfillment, humanistic psychologists have also published papers on a wide range of social issues and topics, such as the promotion of international peace and understanding, awareness of the holocaust, the reduction of violence, and the promotion of social welfare and justice for all.

Humanistic Psychology has been criticised because its theories are impossible to falsify (Popper, 1969) and lacks predictive power and therefore is not a science. The attempt of many humanistic and positive psychologies to explain all of human behaviour often means that these theories can actually never be proved wrong, which does not mean that they are correct. For instance the psychology of Adler could describe almost any action as a sign that an individual has overcome their feelings of inferiority or alternatively that same behaviour could be described as a failure in this respect. These theories are the scientific equivalent of saying 'either it is raining or it is not'. A good scientific theory should be able be falsifiable and have predictive power (Chalmers, 1999); thefore humanistic psychology is not a science.

PSYCHOANALYSIS

Psychoanalysis is a family of psychological theories and methods based on the work of Sigmund Freud. As a technique of psychotherapy, psychoanalysis seeks to discover connections among the unconscious components of patients' mental processes. The analyst's goal is to help liberate the patient from unexamined or unconscious barriers of transference and resistance, that is, past patterns of relating that are no longer serviceable or that inhibit freedom.

Origins

Psychoanalysis was devised in Vienna in the 1890s by Sigmund Freud, a neurologist interested in finding an effective treatment for patients with neurotic or hysterical symptoms. As a result of talking with these patients, Freud came to believe that their problems stemmed from culturally unacceptable, thus repressed and unconscious, desires and fantasies of a sexual nature. As his theory developed, Freud developed and cast aside a myriad of different frameworks to model and explain the phenomena he encountered in treating his patients.

Schools

Prominent current schools of psychoanalysis include:

- Self psychology, which emphasizes the development of a stable sense of self through mutually empathic contacts with other humans

- Lacanian psychoanalysis, which integrates psychoanalysis with semeiotics and Hegelian philosophy
- Analytical psychology, which has a more spiritual approach, founded by Carl Jung
- Object relations theory, which stresses the dynamics of one's relationships with internal, fantasized, others; founded by Melanie Klein
- Interpersonal psychoanalysis, which accents the nuances of interpersonal interactions
- Relational psychoanalysis, which combines interpersonal psychoanalysis with object-relations theory.
- Modern psychoanalysis, a body of theoretical and clinical knowledge developed by Hyman Spotnitz and his colleagues, extended Freud's theories so as to make them applicable to the full spectrum of emotional disorders. Modern psychoanalytic interventions are primarily intended to provide an emotional-maturational communication to the patient, rather than to promote intellectual insight.

Although these schools have dramatically different theories, most of them continue to stress the strong influence of self-deception and the influence a person's past has on their current mental life. Today psychoanalytic ideas are embedded in the culture, especially in childcare, education, literary criticism, and in psychiatry, particularly medical and non-medical psychotherapy. Though there is a mainstream of evolved analytic ideas, there are groups who more specifically follow the precepts of one or more of the later theoreticians.

Technique

The basic method of psychoanalysis is the transference and resistance analysis of free association. The patient, in a relaxed posture, is directed to say whatever comes to mind. Dreams, hopes, wishes, and fantasies are of interest, as are recollections of early family life. Generally the analyst simply listens, making comments only when, in his or her professional judgment, an opportunity for insight on the part of the patient arises. In listening, the analyst attempts to maintain an attitude of empathic neutrality, a nonjudgmental stance designed to create a safe environment. The analyst asks that the analysand (patient) speak with utter honesty about whatever comes to awareness while interpreting the patterns and inhibitions that appear in the patient's speech and other behavior.

Many clinicians hold that psychoanalysis is not recommended in cases of serious psychological disruption, such as psychosis, suicidal depression, or severe untreated alcoholism. Such patients may be labeled "un-analyzable". More typical applications include treatment of clinical depression and personality disorders.

Some more recent forms of psychoanalysis seek to help patients gain self-esteem through greater trust of the self, overcome the fear of death and its effects on current behavior, and maintain several relationships that appear to be incompatible.

Although single-client sessions remain the norm, psychoanalysis has been adapted as a form of group therapy by Harry Stack Sullivan and others.

Efficacy

Currently, most psychoanalysts claim that analysis is most useful as a method in cases of neurosis and with character or personality problems. Psychoanalysis is believed to be most useful in dealing with ingrained problems of intimacy and relationship and for those problems in which established patterns of life are problematic. As a therapeutic treatment, psychoanalysis generally takes three to five meetings a week and requires the amount of time for natural or normal maturational change (three to seven years).

Analysis of previous randomized controlled trials have suggested that psychoanalytic treatment is more effective than the absence of treatment in specific psychiatric disorders.. Empirical research on the efficacy of psychoanalysis and psychoanalytic psychotherapy has also become prominent among psychoanalytic researchers.

Research on psychodynamic treatment of some populations shows mixed results. Research by analysts such as Bertram

Karon and colleagues at Michigan State University had suggested that when trained properly, psychodynamic therapists can be effective with schizophrenic patients. More recent research casts doubt on these claims. The Schizophrenia Patient Outcomes Research Team (PORT) report argues in its Recommendaton 22 against the use of psychodynamic therapy in cases of schizophrenia, noting that more trials are necessary to verify its effectiveness. However, the PORT recommendation is based on the opinions of clinicians rather than on empirical data, and empirical data exist that contradict this recommendation (link to abstract). A review of current medical literature in The Cochrane Library, (the updated abstract of which is available online) reached the conclusion that no data exist that demonstrate that psychodynamic psychotherapy is effective in treating schizophrenia. Further data also suggest that psychoanalysis is not effective (and possibly even detrimental) in the treatment of sex offenders.

Cost and Length

The cost of psychoanalytic treatment ranges widely, from as low as ten dollars a session (with an analytic candidate in training at an institute) to over 250 dollars a session with a senior training analyst.

Length of treatment varies. Some psychodynamic approaches, such as Brief Relational Therapy (BRT), Brief Psychodynamic Therapy (BPT), and Time-Limited Dynamic Therapy (TLDP) limit treatment to 20-30 sessions. Full-fledged psychoanalysis, however, generally lasts longer, with an average of 5.7 years, according to a recent survey. Which treatment length is optimal depends on the individual's needs. Managed care has placed increased pressure on psychotherapy in general to restrict time devoted to patients.

Training

Throughout the history of psychoanalysis, most psychoanalytic organizations have existed outside of the university setting, with a few notable exceptions.

Psychoanalytic training usually occurs at a psychoanalytic institute and may last approximately 4-10 years. Training includes coursework, supervised psychoanalytic treatment of patients, and personal psychoanalysis lasting 4 or more years.

An ongoing debate in professional psychoanalysis concerns the qualifications candidates must have to enter analytic training. Freud believed that applicants from the humanities and many nonmedical disciplines are as well prepared as physicians for psychoanalytic training.

Anna Freud described the essential personal qualities in a psychoanalyst being the following.

The American Psychoanalytic Association, however, limited access to training to medical doctors until quite recently. Later, after extensive debates and legal battles, psychoanalytic training in most institutes was opened to nonmedical mental health professionals, such as psychologists and clinical social workers. Currently, access to training by applicants from nonclinical disciplines, such as literary studies and philosophy, is limited to a handful of institutes in the United States. By contrast, many or most institutes in Europe and Latin America admit candidates without formal clinical training.

Although the popularity of psychoanalysis was in decline during the 1980s and early 1990s, prominent psychoanalytic institutes have experienced an increase in the number of applicants in recent years.

Distinctive Theories

The theories distinctive of psychoanalysis generally include the following hypotheses:

- Human development is best understood in terms of changing objects of sexual desire.
- The psychic apparatus habitually represses wishes, usually of a sexual or aggressive nature, whereby they become preserved in one or more unconscious systems of ideas.
- Unconscious conflicts over repressed wishes have a tendency to manifest themselves in dreams, parapraxes ("Freudian slips"), and symptoms.
- Unconscious conflicts are the source of neuroses.

- Neuroses can be treated through bringing the unconscious wishes and repressed memories to consciousness in psychoanalytic treatment.

The Unconscious and Psychic Structures

The unconscious refers to that part of mental functioning of which subjects make themselves unaware. The psychoanalytic unconscious is similar to but not precisely the same as the popular notion of the subconscious. For psychoanalysis, the unconscious does not include all of what is not conscious. It does not include e.g., motor skills, but only what is actively repressed from conscious thought. It does include instances of automatic processing such as stereotypes and the effects of past relationships on the present.

For Freud, the unconscious was a depository for socially unacceptable ideas, wishes or desires, traumatic memories, and painful emotions put out of mind by the mechanism of psychological repression. However, the contents did not necessarily have to be solely negative. In the psychoanalytic view, the unconscious is a force that can only be recognized by its effects-it expresses itself in the symptom.

The ego, super-ego, and id are the divisions of the psyche according to Freud's later "structural theory". The id contains "primitive desires" (hunger, rage, and sex), the super-ego contains internalized norms, morality and taboos, and the ego mediates between the two and may include or give rise to the sense of self.

Roots of Neurosis

In his earliest writings on the subject, Freud theorized that all neuroses were rooted in childhood sexual abuse (the so-called seduction theory). Later, Freud çame to abandon or de-emphasize this hypothesis, emphasizing instead the importance of unconscious fantasy as the cause of neurosis, particularly fantasy structured according to the Oedipus complex.

The Oedipus complex is a concept developed by Sigmund Freud to explain the origin of certain neuroses in childhood. It is defined as a male child's unconscious desire for the exclusive love of his mother. This desire includes jealousy towards the father and the unconscious wish for that parent's death. The idea is based on the Greek myth of Oedipus, who unwittingly kills his father Laius and marries his mother Jocasta.

Freud revisited the Oedipal territory in the final essay of Totem and Taboo. There, he combined one of Charles Darwin's more speculative theories about the arrangements of early human societies (a single alpha-male surrounded by a harem of females, similar to the arrangement of gorilla groupings) with the theory of the sacrifice ritual taken from William Robertson Smith. Smith believed he had located the origins of totemism in a singular event, whereby a band of prehistoric brothers expelled from the alpha-male group returned to kill their father, whom they both feared and respected. In this respect, Freud located the beginnings of the Oedipus complex at the origins of human society, and postulated that all religion was in effect an extended and collective form of guilt and ambivalence to cope with the killing of the father figure (which he saw as the true original sin).

The Life and Death Instincts

Freud believed that humans were driven by two conflicting central desires: the life drive (Eros) (incorporating the sex drive) and the death drive. Freud's description of Eros/Libido/ included all creative, life-producing drives. The death drive (or death instinct) represented an urge inherent in all living things to return to a state of calm, or, ultimately, of non-existence. The presence of the death drive was only recognized in his later years, and the contrast between the two represents a revolution in his manner of thinking.

Post-Freudian Schools

- Object relations theory is the idea that the ego-self exists only in relation to other objects, which may be external or internal. Internal objects are internalized versions of external objects, primarily formed from early interactions with the parents. According to object-relations theory there are three fundamental "affects" that can exist between the self and the other-attachment, frustration, and rejection. These affects are alleged to be universal emotional states that are major building blocks of the personality. Object relations theory was pioneered in the 1920s by the Austrian psychoanalyst Otto Rank, in the 1930s by Melanie Klein and in the 1940s and 1950s by Ronald Fairbairn, D.W. Winnicott, Harry Guntrip, Margaret Mahler and others.

Interpersonal psychoanalysis is based on the theories of Harry Stack Sullivan, an American psychiatrist who believed that the details of patient's interpersonal interactions with others provided insight into the causes and cures of mental disorder. Sullivan argued that patients keep many aspects of interpersonal relationships out of their awareness by selective inattention. He felt it to be important for psychotherapists to conduct a detailed inquiry into patient's interactions with others so that patients would become optimally aware of their interpersonal patterns.

Modern Adaptations

Cultural Adaptations

Psychoanalysis can be adapted to different cultures, as long as the therapist or counseling understands the client's culture. For example, Tori and Blimes found that defense mechanisms were valid in a normative sample of 2,624 Thais. The use of certain defense mechanisms was related to cultural values. For example Thais value calmness and collectiveness (because of Buddhist beliefs), so they were low on regressive emotionality. Psychoanalysis also applies because Freud used techniques that allowed him to get the subjective perceptions of his patients. He takes an objective approach by not facing his clients during his talk therapy sessions. He met with his patients where ever they were, such as when he used free association-where clients would say whatever came to mind without self-censorship. His treatments had little to no structure for most cultures, especially Asian cultures. Therefore, it is more likely that Freudian constructs will be used in structured therapy (Thompson, et al., 2004). In additi n, Corey postulates that it will be necessary for therapist to help clients develop a cultural identity as well as an ego identity. Since Freud has been criticized for not accounting for external/societal forces, it seems logical that therapists or counselors using his premises will work with the family more. Psychoanalytic constructs fit with constructs of other more structured therapies, and Firestone (2002) thinks psychotherapy should have more depth and involve both psychodynamic and cognitive-behavioral approaches. For example, Corey states, that Ellis, the founder of Rational Emotive Behavioral Therapy (REBT) would allow his clients to experience depression over a loss, such an emotion would be rational-often people will be irrational and deny their feelings.

Adaptations for Age and Managed Care

Play Therapy for Different Ages

Psychoanalytic constructs can be adapted and modified to both age and managed care through the use of play therapy such as art therapy, creative writing, storytelling, bibliotherapy, and psychodrama. In the 1920's, Anna Freud (Sigmund Freud's daughter) adapted psychoanalysis for children through play. Using toys and games, she was able to enhance relationship with the child-Freud has been criticized for his, objective and disengaged, approach. When children play, they often engage in a make believe world where they can express their fears and fantasies, and they do so without censorship, so it resembles very much the technique of free association. Psychoanalytic play therapy allows the child and the counselor to access material in the unconscious, material that was avoided and forgotten. This material is re-integrated into the conscience, and the counselor is able to work with the child and the family to address the trauma or issue that was forgotten. With adults, the term art therapy is used, instead of play, however they are synonymous. The counselor simply adapts art therapy to the age of the client. With children, a counselor may have a child draw a portrait of his self, and then tell a story about the portrait. The counselor watches for re-occurring themes-regardless of whether it is with art or toys. With adults, the counselor may work one on one or in a group and have clients do various art activities like painting or clay to express themselves-toys here would not probably not be age appropriate, and children stop pretend play as they transition into adolescence. Since play is considered appropriate in Occidental (Western) culture, it allows people to deal with personal/social issues that they would normally avoid-it allows them to drop their defenses without anxiety and fear.

Other Play Therapy Techniques

Bibliocounseling involves selecting stories from books that children can identify with (similar issues). Through this story, a child will be more likely to not feel defensive and will work to find alternative solutions to problems. Storytelling is similar, the counselor may tell a story but not use a name, and instead he may address the child with each new sentence using his name. For example, He may say, "next, Eric, the little boy had dream about a mouse that was not like the other mice..." Play therapy for Managed Care Unlike traditional psychoanalysis, play therapy takes much shorter time span; which allow insurance companies to cover it for their clients. Even more, it provides more structure to the process allowing for specific measurable goals. Psychoanalytic theory will be applied in more preventative ways, such as educating parents on how to best meet the needs of the child and enhance the child's development and growth. Lastly, more advocates may use homework assignments such as journal writing to save time (Thompson et al., 2004). Expressive Writing for Managed Care According to a book, review by Berman (2003) the writing cure provides an analysis of research that supports expressive writing as a way to integrate cognitions and work through trauma. People who write about traumatic events experience more self control. The Writing Cure offers new, cost-effective ways to treat clients; clients can even use expressive writing to work through their own personal/social issues.

Criticisms

Psychoanalysis has been criticized on a variety of grounds by Karl Popper, Adolf Grünbaum, Mario Bunge, Hans Eysenck, Ludwig Wittgenstein, Hans Israels, Richard Webster, Peter Medawar, Ernest Gellner, Frank Cioffi, Frederick Crews, Arthur Janov, William Sargant and others. Popper argues that it is not scientific because it is not falsifiable. Grünbaum argues that it is falsifiable, and in fact turns out to be false. Janov claims that psychoanalysis intellectualizes the feeling process which only strengthens defenses. The other schools of psychology have produced alternative methods of psychotherapy, including behavior therapy, cognitive therapy, Gestalt therapy, primal therapy and person-centered psychotherapy. Exchanges between critics and defenders of psychoanalysis have often been so heated that they have come to be characterized as the Freud Wars.

An important consequence of the wide variety of psychoanalytic theories is that different schools of psychoanalysis criticise each other. One consequence is that some critics offer criticism of specific ideas present only in one or more theories, rather than in all of psychoanalysis while not rejecting other premises of psychoanalysis. Defender of psychoanalysis argue that many critics (such as feminist critics of Freud) have attempted to offer criticisms of psychoanalysis that were in fact only criticisms of specific ideas present only in one or more theories, rather than in all of psychoanalysis. As the psychoanalytic researcher Drew Westen puts it, "Critics have typically focused on a version of psychoanalytic theory-circa 1920 at best-that few contemporary analysts find compelling...In so doing, however, they have set the terms of the public debate and have led many analysts, I believe mistakenly, down an indefensible path of trying to defend a 75 to 100-year-old version of a theory and therapy that has changed substantially since Freud laid its foundations at the turn of the century." link to Westen article On the other hand, those who criticise psychoanalysis on scientific basis tend to dismiss the entire subject as pseudo science.

Scientific Validity

An early and important criticism of psychoanalysis was that its theories were based on little quantitative and experimental research, and instead relied almost exclusively on the clinical case study method. In comparison, brief psychotherapy approaches such as behavior therapy and cognitive therapy have shown much more concern for empirical validation (Morley et al. 1999). Some even accused Freud of fabrication including the famous and miraculous cure of Anna O. (Borch-Jacobsen 1996).

An increasing amount of psychoanalytic research from academic psychologists and psychiatrists who have worked to quantify and measure psychoanalytic concepts has begun to address this criticism. However, a survey of scientific research showed that while personality traits corresponding to Freud's oral, anal, Oedipal, and genital phases can be observed, they cannot be observed as stages in the development of children, nor it be confirmed that such traits in adults result from childhood experiences (Fisher & Greenberg, 1977, p399).

However, these stages should not be viewed as crucial to modern psychoanalysis. What is crucial to modern psychoanalytic theory and practice is the power of the unconscious and the transference phenomenon. Some claim the idea of "unconscious" is contested because human behaviour can be observed while human psychology have to be guessed at. However, the unconscious is now a hot topic of study at the undergraduate and graduate level in the fields of experimental and social psychology (e.g., implicit attitude measures, fMRI, and PET scans, and other indirect tests). One would be hard pressed to find scientists who still think of the mind as a "black box." Presently, the field of psychology has embraced the study of things outside one's awareness. Even strict behaviorists acknowledge that a vast amount of classical conditioning is unconscious and that this has profound effects on our emotional life. The idea of unconscious and the transference phenomenon have been widely researched and, it is claimed, to be validated in the fields of cognitive psychology and social psychology, though such claims are also contested. The recent development in neuroscience resulted in one side arguing that it have provided a biological basis for unconscious emotional processing in line with psychoanalytic theory while the other side arguing that such finding make psychoanalytic theory obsolte and irrelevant.

E. Fuller Torrey, considered by some to be a leading American psychiatrist, writing in Witchdoctors and Psychiatrists (1986) stated that psychoanalytic theories have no more scientific basis than the theories of traditional native healers, "witchdoctors" or modern "cult" alternatives such as est. In fact, an increasing number of scientists regard psychoanalysis as a pseudoscience (Cioffi, F. 1998).

Among philosophers, Karl Popper argued that Freud's theory of the unconscious was not falsifiable and therefore not scientific. Popper did not object to the idea that some mental processess could be unconscious but to investigations of the mind that were not falsifiable. In other words, if it were possible to connect every conceivable experimental outcome with Freud's theory of the unconscious mind, then no experiment could refute the theory.

Some proponents of psychoanalysis suggest that its concepts and theories are more akin to those found in the humanities than those proper to the physical and biological/medical sciences, though Freud himself tried to base his clinical formulations on a hypothetical neurophysiology of energy transformations. For example, the philosopher Paul Ricoeur argued that psychoanalysis can be considered a type of textual interpretation or hermeneutics. Like cultural critics and literary scholars, Ricoeur contended, psychoanalysts spend their time interpreting the nuances of language-the language of their patients. Ricoeur claimed that psychoanalysis emphasizes the polyvocal or many-voiced qualities of language, focusing on utterances that mean more than one thing. Ricoeur classified psychoanalysis as a hermeneutics of suspicion. By this he meant that psychoanalysis searches for deception in language, and thereby destabilizes our usual reliance on clear, obvious meanings.

Controversy over Efficacy

Psychoanalysts, for most of their history, have relied on the clinical case report as the chief method of evaluating the efficacy of treatment. In comparison, brief psychotherapy approaches such as behavior therapy and cognitive therapy have shown much more concern for empirical validation from broad-based studies modeled after randomized pharmaceutical trials.

At least in the United States, psychoanalysis has usually been perceived as a form of insight-based therapy, with the goal of bringing unconscious thoughts or memories into consciousness. Some studies, however, throw doubt on whether insight is a necessary or sufficient means for improving a person's behavior or increasing their level of functioning (Fisher & Greenberg, 1977, pp411-412).

There is even a great controversy among psychologists as to whether repressed memories actually exist. The whole recovered memory movement, which has flourished in the United States in the last decade, is now highly criticized by the advocates of the false memory syndrome (Loftus & Ketcham, 1994).

Theoretical Criticism

Psychoanalysts have often complained about the significant lack of theoretical agreement among analysts of different schools. Many authors have attempted to integrate the various theories, with limited success. However, with the publication of the Psychodynamic Diagnostic Manual much of this lack of cohesion has been resolved.

The philosopher Jacques Derrida incorporated certain aspects of psychoanalytic theory into his practice of deconstruction in order to question what he called the 'metaphysics of presence' or 'self-presence'. This was the defining trait (for Derrida) of traditional metaphysics, namely its assumption that the meaning of utterances can be pinned down and made fully evident to consciousness, perhaps most evident in Descartes' conception of 'clear and distinct ideas'. Derrida is here influenced by Freud (among others such as Marx and Nietzsche.) For instance, Freud's insistence, in the first chapter of The Ego and the Id, that philosophers will recoil from his theory of the unconscious is clearly a forbearer to Derrida's understanding of metaphysical 'self-presence'. However, Derrida goes on to turn certain of these practices against Freud himself, in order (in Derrida's typical manner) to reveal tensions and contradictions in Freud's work which are nonetheless the very conditions upon which it can operate-its simultaneous conditions of possibility and impossibility.

For instance, although Freud will define religion and metaphysics as a displacement of the identification with the father in the resolution of the Oedipal complex (e.g. in 'The Ego and The Id' and 'Totem and Taboo') Derrida will insist (for instance in 'The Postcard') that the prominence of the father in Freud's own analysis is at the same time indebted to and an example of the prominence given to the father in Western metaphysics and theology since Plato. Thus, (in a similar manner to that in which Levi-Strauss reads Freud's understanding of the Oedipal complex as but another version of the Oedipus myth,) Derrida understands Freud as remaining partly within that theologico-metaphysical tradition ('phallologocentrism' Derrida helpfully calls it) which Freud nonetheless criticizes. However, the purpose of Derrida's analysis is not to refute Freud per se, (which would only be to reaffirm traditional metaphysics) but rather to reveal an aporia (an undecidability) at the very heart of Freud's project. Such a 'deconstruction' (or indeed psychoanalysis) of Freud does tend to cast doubt upon the possibilty of delimiting psychoanalysis as a rigorous science. However in doing so it celebrates and pledges a critical alliegance to that side of Freud which emphasises the open-ended and improvisatory nature of psychoanalysis, and its (methodical and ethical) demand (for instance in the opening chapters of the 'Interpretaion of Dreams,') that the testimony of the analysand should be given prominence in the practice of analysis.

Psychoanalysis, or at least the dominant version of it, has been denounced as patriarchal or phallocentric by proponents of feminist theory. Other feminist scholars appreciate how Freud opened up society to female sexuality.

PSYCHODYNAMIC PSYCHOTHERAPY

Psychodynamic psychotherapy is a type of psychotherapy, where the psychotherapist and client meet about once or twice a week. It is different from other systems of psychotherapy, for instance psychoanalysis or cognitive therapy in that it uses a range of different techniques, applied to the client considering his or her needs. A psychodynamic therapist may find that Object relations theory may be best for a client with Borderline Personality Disorder, and the next client who displays some anxiety in her marriage may be given some cognitive therapy to give symptom relief.

History

Psychodynamics was born with the 1874 publication of Lectures on Physiology by German scientist Ernst von Brucke who supposed that all living organisms are energy-systems governed by the principle of the conservation of energy. During this year, at the University of Vienna, Brucke was also coincidentally the supervisor for first-year medical student Sigmund Freud who adopted this new "dynamic" physiology. Later, the theory of psychodynamics was developed further by those such as Carl Jung, Alfred Adler, and Melanie Klein.

Approaches

Most psychodynamic approaches are centered around the idea of a maladapted function developed early in life (usually childhood) which are at least in part unconscious. This maladapted function (a.k.a. defense mechanism) does not do well as it formed instead of a normal/healthy one. Later on the client will feel discomfort when they notice (or do not notice) that this function causes problems day to day. The psychodynamic therapist will first treat the discomfort associated with the poorly formed function, reveal to the client that such a function exists, then change, remove or replace it with a proper one.

Psychodynamic psychotherapy involves a great deal of introspection and reflection from the client. Usually this level of insight is unfettered when the client wants to be helped or is pushed by family or friends. Speaking to this is also the client's ability to dive into their past; they must possess enough resilience and ego-strength to deal with/use the onslaught of feeling a new perspective brings. The more fragile client may be treated with a different treatment, for instance, cognitive therapy.

Objective

The "goal" of psychodynamic therapy is the experience of "truth." This "truth" must be encountered through the breakdown of psychological defenses. Simply stated: psychotherapy teaches the client to be honest. Of course, this requires considerable time, money, and effort. Individuals suffering from "psychological disorders" or deep-rooted "personality disorders," often come from confusing, manipulative, dishonest, or even violent families in childhood. Being honest with ones "feelings" is a difficult, even terrifying process for these people. But there is a silver lining. If the patient is willing to face up to their hidden secrets: they will discover the unconscious "reason" for many of their feelings, and therefore obtain self-understanding and relief. In essence, the more honest and direct one is with his/her life, the more "symptoms" will dissolve, and the more one's childhood and defenses are understood.

TRANSPERSONAL PSYCHOLOGY

Transpersonal psychology is a school of psychology that studies the transcendent, or spiritual dimensions of humanity. Among these factors we find such issues as self-development, peak experiences, mystical experiences and the possibility of development beyond traditional ego-boundaries. Thus the interest in human experiences which apparently are 'trans-personal,'or 'trans-egoic'. A short definition from the Journal of Transpersonal Psychology suggests that transpersonal psychology is concerned with the study of humanity's highest potential, and with the recognition, understanding, and realization of unitive, spiritual, and transcendent states of consciousness (Lajoie and Shapiro, 1992:91).

The field is considered by proponents to be the "fourth force"' in the field of psychology, the three other fields being psychoanalysis, behaviorism, and humanistic psychology. According to transpersonal theory, these other schools of psychology have failed to give weight to transpersonal or "transegoic" elements of human existence, such as religious conversion, altered states of consciousness, trance and spirituality, in their academic reflection. Thus, transpersonal psychology strives to combine insights from modern psychology with insights from the world's contemplative traditions, both East and West. The transpersonal and spiritual dimensions of the psyche have traditionally not been a focus of interest for Western psychology, which has mainly focused on the prepersonal and personal aspects of the human psyche (Cowley & Derezotes, 1994; Miller, 1998).

Definitions of Transpersonal Psychology

Lajoie and Shapiro (1992) reviewed forty definitions of transpersonal psychology that had appeared in literature over the period 1969 to 1991. They found that five key themes in particular featured prominently in these definitions: states of consciousness, higher or ultimate potential, beyond the ego or personal self, transcendence and the spiritual. Walsh and Vaughan (1993) have criticised many definitions of transpersonal psychology, for carrying implicit ontological or methodological assumptions. They also challenge definitions that link transpersonal psychology to healthy states only, or to the Perennial Philosophy. These authors define transpersonal psychology as being the branch of psychology that is concerned with transpersonal experiences and related phenomena, noting that "These phenomena include the causes, effects and correlates of transpersonal experiences, as well as the disciplines and practices inspired by them" (Walsh & Vaughan, 1993, p203).

The Development of the Field

Among the thinkers who are considered to have set the stage for transpersonal studies are William James, Sigmund Freud, Otto Rank, Carl Jung, Abraham Maslow, and Roberto Assagioli (Cowley & Derezotes, 1994; Miller, 1998; Davis, 2003). Research by Vich (1988) suggests that earliest usage of the term "transpersonal" can be found in lecture notes which William James had prepared for a semester at Harvard University in 1905-6. A major motivating factor behind the initiative to establish this school of psychology was Abraham Maslow's already published work regarding human peak experiences. Maslow's work grew out of the humanistic movement of the 1960's, and gradually the term "transpersonal" was associated with a distinct school of psychology within the humanistic movement.

In 1969, Abraham Maslow, Stanislav Grof and Anthony Sutich were the initiators behind the publication of the first issue of the Journal of Transpersonal Psychology, the leading academic journal in the field. This was soon to be followed by the founding of the Association for Transpersonal Psychology (ATP) in 1972. Past presidents of the association include Alyce Green, James Fadiman, Frances Vaughan, Arthur Hastings, Daniel Goleman, Robert Frager, Ronald Jue, Jeanne Achterberg and Dwight Judy. In the 1980s and 90s the field developed through the works of such authors as Stanislav Grof, Ken

Wilber, Michael Washburn, Frances Vaughan, Roger Walsh, Stanley Krippner, Michael Murphy, Charles Tart, David Lukoff and Stuart Sovatsky. While Wilber has been considered an influential writer and theoretician in the field, he has since personally dissociated himself from the movement in favor of what he calls an integral approach.

Today transpersonal psychology also includes approaches to health, social sciences and practical arts. Transpersonal perspectives are also being applied to such diverse fields as psychology, psychiatry, anthropology, sociology, pharmacology, cross-cultural studies (Scotton, Chinen and Battista, 1996; Davis, 2003) and social work (Cowley & Derezotes, 1994). Currently, transpersonal psychology, especially the schools of Jungian and Archetypal psychology, is integrated, at least to some extent, into many psychology departments in US and European Universities. Transpersonal therapies are also included in many therapeutic practices.

Institutions of higher learning that have adopted insights from Transpersonal psychology include The Institute of Transpersonal Psychology (US), John F. Kennedy University (US), California Institute of Integral Studies (US), Saybrook Institute (US), Naropa University (US), the University of Northampton (UK) and Liverpool John Moores University (UK). There is also a strong connection between the transpersonal and the humanistic perspective. This is not surprising since transpersonal psychology started off within humanistic psychology (Aanstoos, Serlin & Greening, 2000).

By common consent, the following branches are considered to be transpersonal psychological schools: Jungian psychology, depth psychology (more recently rephrased as Archetypal psychology by James Hillman), spiritual psychology (Robert Sardello), psychosynthesis founded by Roberto Assagioli, and the theories of Abraham Maslow, Stanislav Grof, Ken Wilber, Michael Washburn and Charles Tart.

Transpersonal psychology is sometimes confused with parapsychology, a mistake made due to the overlapping and unconventional research interests of both fields; parapsychology would however tend to focus more in its subject matter on the "psychic" and transpersonal psychology the "spiritual" (relatively crude though these categorizations are, it is still a useful distinction in this context). While parapsychology leans more towards traditional scientific epistemology (laboratory experiments, statistics, research on cognitive states), transpersonal psychology tends to be more closely related to the epistemology of the humanities and the hermeneutic disciplines (humanism, existentialism, phenomenology, anthropology), although it has always included contributions involving experimental and statistical research.

Transpersonal psychology is also sometimes confused with the New Age. Although the transpersonal perspective grew out of the human potential movement, a movement that many commentators associate with a broad conception of the New Age, it is still problematic to place Transpersonal Psychology within such a framework. Transpersonal psychology is an academic discipline, not a religious or spiritual movement, and many of the field's leading authors, among those Sovatsky (1998) and Rowan (1993), have addressed problematic aspects of New Age hermeneutics. Assocations between transpersonal psychology and the New Age have probably contributed to the failures in the U.S. to get transpersonal psychology more formally recognised within the professional body, the American Psychological Association (APA). A significant breakthrough in this context was the successful establishment of a Transpersonal Psychology Section within the British Psychological Society (the UK professional body equivalent to the APA) in 1996, co-founded by David Fontana, Ingrid Slack and Martin Treacy, "the first Section of its kind in a Western scientific society" according to Fontana (Fontana et. al, 2005, p.5).

Criticisms of Transpersonal Psychology

Criticisms of transpersonal psychology have come from several commentators. One of the earliest criticisms of the field was issued by the Humanistic psychologist Rollo May, who disputed the conceptual foundations of transpersonal psychology (Aanstos, Serling & Greening, 2000). May was particularly concerned about the low level of reflection on the dark side of human nature, and on human suffering, among the early transpersonal theorists. A similar critique was also put forward by Alexander (1980) who thought that Transpersonal

Psychology, in light of the thinking of William James, represented a philosophy that failed to take evil adequately into account. This serious criticism has been absorbed by later Transpersonal theory, which has been more willing to reflect on these important dimensions of human existence (Scotton, Chinen and Battista, 1996; Daniels, 2005). Criticism has also come from the cognitive psychologist, and humanist, Albert Ellis (1989) who has questioned transpersonal psychology's scientific status and its relationship to religion and mysticism. This criticism has been answered by Wilber (1989) and Walsh (1989).

Friedman (2000) has criticized the field of Transpersonal psychology for being underdeveloped as a field of science, and he mentions a number of factors which he believe are anti-scientific beliefs prevalent in the field. He further differentiates between Transpersonal psychology as a field of scientific psychology, and the larger area of transpersonal studies which, according to the author, may include a number of unscientific approaches. Doctrines or ideas of many colorful personalities, who were or are spiritual teachers in the Western world, such as Gurdjieff or Alice Bailey, are often assimilated in the transpersonal psychology mainstream scene. This development is, generally, seen as detrimental to the aspiration of transpersonal psychologists to gain a firm and respectable academic status. However, Scotton, Chinen and Battista (1996) believe that much of this criticism can be nuanced if one differentiates between the field of Transpersonal Psychology on the one hand, and a popular mainstream scene that operates outside of an academic context, on the other.

It could also be argued that most psychologists do not hold strictly to traditional schools of psychology-most psychologists take an eclectic approach. This could mean that the transpersonal categories listed are considered by standard subdisciplines of psychology; religious conversion falling within the ambit of social psychology, altered states of consciousness within physiological psychology, and spiritual life within the psychology of religion. Transpersonal psychologists, however, disagree with the approach to such phenomena taken by traditional psychology, and claim that transpersonal categories have typically been dismissed either as signs of various kinds of mental illnesses, or as a regression to infantile stages of psychosomatic development. Thus, as illustrated by the pre/ trans fallacy, religious and spiritual experiences have in the past been seen as either regressive or pathological and treated as such.

PSYCHOLOGY OF RELIGION

U.S. psychologist and philosopher William James (1842-1910) is regarded by most psychologists of religion as the founder of the field. He served as president of the American Psychological Association, and wrote one of the first psychology textbooks. In the psychology of religion, James' influence endures. His Varieties of Religious Experience is considered to be the classic work in the field, and is worth reading for anyone interested in psychology and religion. Indeed, references to James' ideas are common at professional conferences.

James distinguished between institutional religion and personal religion. Institutional religion refers to the religious group or organization, and plays an important part in a society's culture. Personal religion, in which the individual has a mystical experience, can be experienced regardless of the culture. James was most interested in understanding personal religious experience. If personal religious experiences were what James preferred, dogmatism was something he disliked. The importance of James to the psychology of religion-and to psychology more generally-is difficult to overstate. He discussed many essential issues that remain of vital concern today.

In studying personal religious experiences, James made a distinction between healthy-minded and sick-souled religiousness. Individuals predisposed to healthy-mindedness tend to ignore the evil in the world and focus on the positive and the good. James used examples of Walt Whitman and the "mindcure" religious movement to illustrate healthy-mindedness in The Varieties of Religious Experience. In contrast, individuals predisposed to having a sick-souled religion are unable to ignore evil and suffering, and need a unifying experience, religious or otherwise, to reconcile good and evil. James included quotations from Leo Tolstoy and John Bunyan to illustrate the sick soul.

William James' hypothesis of pragmatism stems from the efficacy of religion. If an individual believes in and performs religious activities, and those actions happen to work, then that practice appears the proper choice for the individual. However, if the processes of religion have little efficacy, then there is no rationality for continuing the practice.

Developmental Approaches to Religion

Attempts have been made to apply stage models, such as that of Jean Piaget and Lawrence Kohlberg, to how children develop ideas about God and about religion in general.

By far the most well-known stage model of spiritual or religious development is that of James W. Fowler, a developmental psychologist at the Candler School of Theology, in his Stages of Faith (ISBN 0-06-062866-9). He follows Piaget and Kohlberg and has proposed a staged development of faith (or spiritual development) across the lifespan in terms of a holistic orientation, and is concerned with the individual's relatedness to the universal.

The book-length study contains a framework and ideas considered by many to be insightful and which have generated a good deal of response from those interested in religion, so it appears to have at least a reasonable degree of face validity. James Fowler proposes six stages of faith development as follows:

1. Intuitive-projective
2. Symbolic Literal
3. Synthetic Conventional
4. Indiduating
5. Paradoxical (conjunctive)
6. Universalising.

Although there is evidence that children up to the age of twelve years do tend to be in the first two of these stages, there is evidence that adults over the age of sixty-one do show considerable variation in displays of qualities of Stages 3 and beyond. Fowler's model has generated some empirical studies, and fuller descriptions of this research (and of these six stages) can be found in Wulff (1991). However, this model has been attacked from a standpoint of scientific research due to methodological weaknesses. Of Fowler's six stages, only the first two found empirical support, and these were heavily based upon Piaget's stages of cognitive development.

The tables and graphs in the book were presented in such a way that the last four stages appeared to be validated, but the requirements of statistical verification of the stages did not come close to having been met. The stuay was not published in a journal, so was not peer-reviewed, and never drew much attention from psychologists.

Other critics of Fowler have questioned whether his ordering of the stages really reflects his own commitment to a rather liberal Christian Protestant outlook, as if to say that people who adopt a similar viewpoint to Fowler are at higher stages of faith development. Nevertheless, the concepts Fowler introduced seemed to hit home with those in the circles of academic religion, and have been an important starting point for various theories and subsequent studies.

Religion and Coping with Stress

Psychologists of religion have looked at how individuals may use religion as a resource in coping with stress. A major contributor here is Kenneth Pargament, whose work shows the influence of attribution theory. Pargament has distinguished styles of coping into the deferring, in which people leave God to see to their problems; the non-religious, in which they do not appeal to God; and the collaborative, in which people believe that a cooperation of God and their own efforts are necessary to help them to cope with stress. Some of Pargament's papers have been published in the Journal for the Scientific Study of Religion.

Evolutionary Psychology of Religion

Evolutionary psychology is based on the presumption that, just like hearts, lungs, livers, kidneys, and immune systems, cognition has functional structure that has a genetic basis, and therefore evolved by natural selection. Like other organs and tissues, this functional structure should be universally shared amongst humans and should solve important problems of survival and reproduction. Evolutionary psychologists seek to understand cognitive processes by understanding the survival and reproductive functions they might serve.

Pascal Boyer is one of the leading figures in the cognitive psychology of religion, a new field of inquiry that is less than fifteen years old, which accounts for the psychological processes that underlie religious thought and practice. In his book Religion Explained, Boyer shows that there is no simple explanation for religious consciousness. Boyer is mainly concerned with explaining the various psychological processes involved in the acquisition and transmission of ideas concerning the gods.

Religious persons acquire religious ideas and practices through social exposure. The child of a Zen Buddhist will not become an evangelical Christian or a Zulu warrior without the relevant cultural experience. While mere exposure does not cause a particular religious outlook (a person may have been raised a Roman Catholic but leave the church), nevertheless some exposure seems required-this person will never invent Roman Catholicism out of thin air.

Boyer claims that cognitive science can help us to understand the psychological mechanisms that account for these manifest correlations and in so doing enable us to better understand the nature of religious belief and practice. To the extent that the mechanisms controlling the acquisitions and transmission of religious concepts rely on human brains, the mechanisms are open to computational analysis. All thought is computationally structured, including religious thought. So presumably, computational approaches can shed light on the nature and scope of religious cognition.

Boyer moves outside the leading currents in mainstream cognitive psychology and suggests that we can use evolutionary biology to unravel the relevant mental architecture. Our brains are, after all, biological objects, and the best naturalistic account of design in nature is Darwin's theory of evolution. To the extent that mental architecture exhibits intricate design, it is plausible to think that the design is the result of evolutionary processes working over vast periods of time. Like all biological systems, the mind is optimised to promote survival and reproduction in the evolutionary environment. On this view all specialised cognitive functions broadly serve those reproductive ends.

For Steven Pinker the universal propensity toward religious belief is a genuine scientific puzzle. He thinks that adaptationist explanations for religion do not meet the criteria for adaptations. An alternative explanation is that religious psychology is a by-product of many parts of the mind that evolved for other purposes.

Religion and Drugs

The American psychologist James H. Leuba (1868-1946), in A Psychological Study of Religion, accounts for mystical experience psychologically and physiologically, pointing to analogies with certain drug-induced experiences. Leuba argued forcibly for a naturalistic treatment of religion, which he considered to be necessary if religious psychology were to be looked at scientifically.

Shamans all over the world and in different cultures have traditionally used drugs, especially psychedelics, for their religious experiences. In these communities the absorption of drugs leads to dreams (visions) through sensory distortion.

William James was also interested in mystical experiences from a drug-induced perspective, leading him to make some experiments with nitrous oxide and even peyote. He concludes that while the revelations of the mystic hold true, they hold true only for the mystic; for others they are certainly ideas to be considered, but hold no claim to truth without personal experience of such.

Drug-induced Religious Experiences

The drugs used by religious communities for their hallucinogenic effects were adopted for explicit and implicit religious functions and purposes. The drugs were and are reported to enhance religious experience through visions and a distortion of the sensory perception (like in dreams in a state of sleep).

- Cannabis, which grows all over the world except in very cold climates, is used in religious practices in Indian and African communities
- Certain Psychedelic mushrooms are used by cultists among the Indians in Latin America, especially in the state of Oaxaca in southern Mexico. The chief species is Psilocybe mexicana, of which the active principles are psilocybin and its derivative psilocin, in their chemical composition and activity not unlike LSD (D-lysergic acid diethylamide); the latter is synthesised from the alkaloids (principally ergotamine and ergonovine) that are constituents of ergot, a growth present in grasses affected by the disease also called ergot. Amanita muscaria (fly agaric) is another mushroom having hallucinogenic properties that has not been thoroughly studied. It may be extremely important, since it may have been the natural source of the ritual soma drink of the ancient Hindus and the comparable haoma used by the Zoroastrians. Fly agaric is mildly toxic at high dosages and is said to have, in addition to its hallucinogenic properties, the ability to increase strength and endurance. It is said also to be a soporific.
- Peyote used by some Indian communities of Mexico. The chief active principle of peyote is an alkaloid called mescaline. Like psilocin and psilocybin, mescaline is reputed to produce visions and other evidences of a mystical nature. Despite claims of missionaries and some government agents that peyote-from the Nahuatl word peyotl ("divine messenger")-is a degenerative and dangerous drug, there appears to be no evidence of this among the members of the Native American Church, a North American Indian cult that uses peyote in its chief religious ceremony. Peyote, like most other hallucinogenic drugs, is not considered to be addictive and, far from being a destructive influence, is reputed by cultists and some observers to promote morality and ethical behaviour among the Indians who use it ritually.

- Ayahuasca, caapi, or yajé, is produced from the stem bark of the vines Banisteriopsis caapi and B. inebrians. Indians who use it claim that its virtues include healing powers and the power to induce clairvoyance, among others. This drink has been certified by investigators to produce remarkable effects, often involving the sensation of flying. The effects are thought to be attributable to the action of harmine, a very stable indole that is the active principle in the plant. While the Indians themselves attribute the properties of the drink Ayahuasca to B. caapi, this is not the common scientific view; the MAOIs present in the B. caapi instead allow the extremely psychedelic ingredients in other plants added to the brew, noticeably plants containing DMT, to be activated and produce an intense experience.
- Kava drink, prepared from the roots of Piper methysticum, a species of pepper, and seemingly more of a hypnotic-narcotic than a hallucinogen, is used both socially and ritually in the South Pacific, especially in Polynesia.
- Iboga, a stimulant and hallucinogen derived from the root bark of the African shrub Tabernanthe iboga is used within the Bwiti religion in Central Africa. The active ingredient in T. iboga is ibogaine, a drug that has been studied for its use in treating addiction.
- Coca, source of cocaine, has had both ritual and social use chiefly in Peru.
- Datura, one species of which is the jimsonweed, is used by native peoples in North and South America; the active principle, however, is highly toxic and dangerous. A drink prepared from the shrub Mimosa hostilis, which is said to produce glorious visions in warriors before battle, is used ritually in the ajuca ceremony of the Jurema cult in eastern Brazil.
- Salvia divinorum, a member of the sage family of plants, is a hallucinogen used by Mazatec shamans for "spiritual journeys" during healing.

The Effects of Meditation

The large variety of meditation techniques shares the common goal of shifting attention away from habitual modes of thinking and perception, in order to permit experiencing in a different way. Many religious and spiritual traditions that employ meditation assert that the world most of us know is an illusion. This illusion is said to be created by our habitual mode of separating, classifying and labelling our perceptual experiences. Meditation is empirical in that it involves direct experience. Though it is also subjective in that the meditative state can be directly known only by the experiencer, and may be difficult or impossible to fully describe in words.

Concentrative meditation can induce an altered state of consciousness characterised by a loss of awareness of extraneous stimuli, one-pointed attention to the meditation object to the exclusion of all other thoughts, and feelings of bliss.

10

Psychology in the Professional World

A psychologist is a scientist and/or clinician who studies psychology, the systematic investigation of the human mind, including behavior and cognition. Psychologists are usually categorized under a number of different fields, the most well-recognized being clinical psychologists, who provide mental health care, and research psychologists, who collect information on people's minds.

In the professional world 'psychologist' has two meanings. In the broadest of these two meanings, psychologist refers to anyone with an advanced degree in clinical psychology, counseling, industrial psychology, educational psychology, or one of several other subfields, and who makes professional contributions based upon that training, be it as a therapist, counselor, researcher, teacher, or consultant. This sense of the word is independent of licensing.

The narrower sense of the word 'psychologist' refers to licensing and to a legal context. In the United States and Canada, 'psychologist' is a protected professional title. In this sense, the title of psychologist means that the mental health professional has a doctoral degree (usually a PhD, PsyD,ScD, or EdD) in clinical, counseling, industrial, or eductional psychology and has also met state or provincial licensing criteria. Those criteria typically include a period of post-doctoral practice under the supervision of a licensed psychologist, a licensing exam, and continuing education requirements. In most states in the United States and in most provinces in Canada, only licensed psychologists, licensed therapists (like Marriage and

Family Therapists) and psychiatrists can legally provide psychotherapy and use this term to refer to aspects of the mental health treatments they perform. Most states exempt from licensing school psychologists who practice within employment by a school district-such psychologists must be certified by their state or province department of education.

Statistics for Licensed Psychologists

These statistics are fiom the United States Department of Labor dated 2004 unless noted otherwise.

- Employed psychologists: 179,000
- 4/10 self-employed
- 1/4 employed by educational institutions (in positions other than teaching)
- Median income for clinical psychologists: US$54,950
- Median income for industrial/organizational psychologists: US$71,400

CLINICAL PSYCHOLOGIST VS PSYCHIATRIST

In the U.S., both licensed psychiatrists and clinical psychologists hold a doctorate in their field. Psychiatrists are physicians who have earned an M.D. or a D.O., whereas psychologists have earned a Ph.D., Psy.D., or Ed.D. Psychiatrists generally spend shorter periods of contact time with clients/ patients, and the principle method of treatment is psychopharmacology. Conversely, clinical psychologists generally rely upon psychological assessment and the use of psychotherapy to relieve psychological distress. It is not uncommon for people suffering from mental illness to combine these services to maximize their impact.

TYPES OF PSYCHOLOGISTS

Psychologists are often categorised under different fields or disciplines.

Clinical Psychologists

Clinical psychologists and counseling psychologists often work in clinics, counseling centers, hospitals and private practices. They diagnose and evaluate mental and emotional disorders, and use tools such as cognitive-behavioral and interpersonal psychotherapy and hypnosis to treat patients. They conduct interviews and psychological tests, and may conduct complex treatment programs, sometimes in conjunction with physicians or other medical specialists. There are specialties within the field of clinical psychology depending on the focus in clinical medicine, including psychopathology, counseling, neuropsychology, medical psychology, clinical health psychology/ behavioral medicine, and forensic psychology.

Forensic Psychologists

Forensic Psychologists are recognized experts in the application of psychological principles to the legal system. Different areas of application include the assessment and treatment of criminal offenders, reporting to courts, working with police, profiling, and advising legal counsel, to name just a few. The majority of forensic psychologists work with offenders, attempting to assess criminogenic needs and identify pathways for their containment.

Medical Psychologists

Medical psychologists are specialists in psychological and behavioral aspects of physical illness and have additional advanced training in psychopharmacology, physiology, and rehabilitation. Their clinical tools include assessment in general and specialist medicine (e.g., chronic illness management, pain treatment, brain injury), clinical psychophysiology (eg, biofeedback), psychotherapy, hypnosis, behavioral medicine, and psychosocial interventions; they are licensed to prescribe medication in certain countries. They tend to work in hospitals and private practice. The specialty is not recognized in all countries.

Health Psychologists

Health psychologists are of two general types: clinical health psychologists (similar to medical psychologists) and community health psychologists (who specialize in psychological and psychosocial aspects of health in the larger community). Clinical health psychologists are oriented to biopsychosocial aspects of physical health and illness and their field overlaps that of behavioral medicine to a large extent. Clinical and community health psychologists work in a number of different roles: clinical diagnosis of problems that relate to physical illness, psychological, psychophysiological, and psychosocial interventions; clinical communication; organisational design in medical care; health promotion; teaching and training; consultancy; giving advice on health policy. The clinical specialty is not recognised independently in all countries (e.g., New Zealand).

School Psychologists

School psychologist work in schools, state departments of education, hospitals, clinics and universities. They are primarily interested in applying psychological knowledge to the resolution of schooling and learning problems. They provide consultation to teachers, administrators, and parents; assessment of students (including assessment of disabilities requiring special education); intervention services; in-service education for staff; family intervention; program evaluation; and research.

Organizational Psychologists

Organizational psychologists are concerned with the performance of people at work and in training, with developing an understanding of how organizations function and how individuals and groups behave at work. Their aim is to increase effectiveness, efficiency and satisfaction at work. This specialty can be highly lucrative.

Research Psychologists

Research psychologists study cognitive and behavioral processes by experimenting on human beings and animals. They work in universities and private research centers, as well as government organizations. Many research psychologists also teach as psychology professors. Common areas of research include perception, memory, motivation, and factors affecting development and social behavior. Unlike clinical and counseling psychologists, research psychologists are generally not licensed.

Professional Licensing Requirements

Licensed psychologists, in virtually every jurisdiction and state in Canada and the U.S., are required to have obtained academic doctoral degrees (typically a Ph.D. or Psy.D.). These academic programs typically take four to six years of postgraduate work to complete and offer some combination of rigorous training in research, clinical practice, and the science of psychology. Licensed psychologists can contribute to scientific research in the form of a dissertation. In this sense, the Ph.D. in professional psychology is a hybrid academic/professional degree (with a greater focus on academics), and university programs in professional psychology are not only academic but are also training programs typically characterized by rigor and intensity. Some psychologists have professional degrees in psychology (Psy.D.), which focus more on clinical practice, and include similarly rigorous coursework, supervised professional training, internship, and developing the ability to conduct and interpret academic research. While only some Psy.D. programs require original scientific research, they typically do require an academic dissertation.

Within Australia the title 'psychologist' is also restricted by law. Use of the title is reserved for individuals registered with their States Psychologist's Registration Board, which requires a four year bachelors degree and either two years of further accredited study or two years of supervised work. These restrictions apply to all who want to use the title 'psychologist' in any of its forms.

In New Zealand, the use of the title 'psychologist' is restricted by law. Initially, only 'clinical psychologist' and 'registered psychologist' were restricted (to people qualified as such). However, in 2004, the use of psychologist is now limited to only those registered psychologists (including clinical psychologists). This is to prevent the misrepresentation of other psychology qualifications in the mental health field. Academic psychologists (e.g., social psychologists) are now only able to refer to themselves as 'researchers in psychology'.

In the UK the use of the title "chartered psychologist" is protected by law. However use of the term 'psychologist', 'psychotherapist' or 'therapist' is not. This can create confusion and varying quality of service as many inadequately trained practitioners who are not chartered can establish practices and charge for services for which they are not properly trained.

Similar restrictions apply in the United States, although application of these restrictions varies state-by-state. Most states include exemptions from licensing in order to use the title "psychologist" if they are teaching in universities, or if they are certified by the state's department of education as a school psychologist and are practicing psychology within the scope of their employment in a school district. In most states, self-employed psychologists offering services to the public, whether they are clinical, counseling, school, or industrial psychologists must be licensed, which usually requires an advanced degree, two years of supervised experience, and passing written and oral examinations. The licensing may carry the title of licensed psychologist, Marriage and Family Therapist, or some other title depending upon the state or province and the license involved.

PRESCRIBING POWERS

In some countries, psychologists cannot obtain a license to prescribe psychiatric medications. Prescription privileges in the United States have recently begun to change, e.g., in New Mexico and Louisiana.

ANALYTICAL PSYCHOLOGY

Analytical psychology (sometimes called Jungian psychology) refers to a school of psychology originating in the ideas of Swiss psychologist Carl Jung and advanced by many other thinkers who followed in his tradition. It is distinct from Freudian psychoanalysis. Its aim is the personal experience of the deep forces and motivations underlying human behavior. Depth psychology and archetypal psychology are related.

An Overview

The overarching goal of Jungian psychology is the reconciliation of the life of the individual with the world of the supra-personal archetypes. Central to this process is the individual's encounter with the unconscious. The human experiences the unconscious through symbols encountered in all aspects of life: in dreams, art, religion, and the symbolic dramas we enact in our relationships and life pursuits. Essential to the encounter with the unconscious, and the reconciliation of the individual's consciousness with this broader world, is learning this symbolic language. Only through attention and openness to this world is the individual able to harmonize their life with these suprapersonal archetypal forces.

"Neurosis" results from a disharmony between the individual's consciousness and the greater archetypal world. The aim of psychotherapy is to assist the individual in reestablishing a healthy relationship to the unconscious (neither being swamped by it-a state characteristic of psychosis-nor completely shut off from it-a state that results in malaise, empty consumerism, narcissism, and a life cut off from deeper meaning). The encounter between consciousness and the symbols arising from the unconscious enriches life and promotes psychological development. Jung considered this process of psychological growth and maturation (which he called the process of individuation) to be of critical importance to the human being, and ultimately to modern society.

In order to undergo the individuation process, the individual must be open to the parts of oneself beyond one's own ego. In order to do this, the modern individual must pay attention to dreams, explore the world of religion and spirituality, and question the assumptions of the operant societal worldview (rather than just blindly living life in accordance with dominant norms and assumptions).

BASIC ASSUMPTIONS

The Unconscious

The basic assumption is that the personal unconscious is a potent part-probably the more active part-of the normal human psyche. Reliable communication between the conscious and unconscious parts of the psyche is necessary for wholeness.

Also crucial is the belief that dreams show ideas, beliefs, and feelings of which individuals may not be readily aware, but need to be, and that such material is expressed in a personalized vocabulary of visual metaphors. Things "known but unknown" are contained in the unconscious, and dreams are one of the main vehicles for the unconscious to express them. Analytical psychology distinguishes between a personal and a collective unconscious.

The collective unconscious contains archetypes common to all human beings. That is, individuation may bring to surface symbols that do not relate to the life experiences of a single person. This content is more easily viewed as answers to the more fundamental questions of humanity: life, death, meaning, happiness, fear. Among these more spiritual concepts may arise and be integrated into the personality.

Self-realization and Neuroticism

An innate need for self-realization leads people to explore and integrate these rejected materials. This natural process is called individuation, or the process of becoming an individual.

According to Jung, Self-realization can be divided into two distinct tiers. In the first half of our lives we separate from humanity. We attempt to create our own identities (I, myself). This is why there is such a need for young men to be destructive and animosity from teens directed at their parents. Jung also said we have a sort of "second puberty" that occurs between 35-40-outlook shifts from emphasis on materialism, sexuality, having kids to concerns about community and spirituality.

In the second half of our lives, we reunite with the human race. We become part of the collective once again. This is when adults start to contribute to humanity (volunteer time, build, garden, create art, etc.) rather than destroy. They are also more likely to pay attention to their unconscious and conscious feelings. How often do you hear a young man state, "I feel angry." or "I feel sad."? This is because they have not rejoined the collective in their older, wiser years, according to Jung. A common theme is for young rebels to "search" for their true selves and realize that a contribution to humanity is essentially a necessity for a whole self.

Jung proposes that the ultimate goal of the collective unconscious and self-realization is to pull us to the highest experience. This, of course, is spiritual. If a person does not proceed toward self-knowledge, neurotic symptoms may arise. Symptoms are widely defined, including, for instance, phobias, fetishism, depression. Symptoms are Self-realization and neuroticism.

Psychoanalysis

Analysis is a way to experience and integrate the unknown material. It is a search for the meaning of behaviors, symptoms, events. Many are the channels to reach this greater self-knowledge. The analysis of dreams is the most common. Others may include expressing feelings in art pieces, poetry or other expressions of creativity. Giving a complete description of the process of dream interpretation and individuation is complex. The nature of the complexity lies on the fact that the process is highly specific to the person who does it.

While Freudian psychoanalysis assumes that the repressed material hidden in the unconscious is given by repressed sexual instincts, Analytical psychology has a more general approach. There is no preconceived assumption about the unconscious material. The unconscious, for Jungian analysts, may contain repressed sexual drives, but also aspirations, fears, etc.

Psychological Type

Analytical psychology distinguishes several psychological types or temperaments.

- Extrovert (Jung's spelling is extrovert, which most dictionaries use; the variant "extravert" is not preferred)
- Introvert

The attitude type could be thought of as the flow of libido (mental energy). The Introvert's flow is inward to the subject and away from the object, ie. external relations. The Extrovert's is outward toward the object, ie. towards external relations and away from the inner, subjective world. Extroverts desire breadth, while introverts seek depth. The Introversion/Extroversion attitude type may also influence mental breakdown. Introverts may be more inclined to catatonic type schizophrenia and extroverts towards manic depression.

The Complex

Early in Jung's career he coined the term and described the concept of the "complex". Jung claims to have discovered the concept during his free association and galvanic skin response experiments. Freud obviously took up this concept in his Oedipus complex amongst others. Jung seemed to see complexes as quite autonomous parts of psychological life. It is almost as if Jung were describing separate personalities within what is considered a single individual. But to equate Jung's use of complexes with something along the lines of multiple personality disorder would be a step out of bounds.

Jung saw an archetype as always being the central organizing structure of a complex. For instance, in a "negative mother complex," the archetype of the "negative mother" would be seen to be central to the identity of that complex. This is to say, our psychological lives are patterned on common human experiences. Interestingly, Jung saw the Ego (which Freud wrote about in German literally as the "I", one's conscious experience of oneself) as a complex. If the "I" is a complex, what might be the archetype that structures it? Jung, and many Jungians, might say "the hero," one who separates from the community to ultimately carry the community further.

Clinical Theories

Jung's writings have been of much interest to people of many backgrounds and interests, including theologians, people from the humanities, and mythologists. Jung often seemed to seek to make contributions to various fields, but he was mostly a practicing psychiatrist, involved during his whole career in treating patients. A description of Jung's clinical relevance is to address the core of his work.

Jung started his career working with hospitalized patients with major mental illnesses, most notably schizophrenia. He was interested in the possibilities of an unknown "brain toxin" that could be the cause of schizophrenia. But the majority and the heart of Jung's clinical career was taken up with what we might call today individual psychodynamic psychotherapy, in gross structure very much in the strain of psychoanalytic practice first formed by Freud.

It is important to state that Jung seemed to often see his work as not a complete psychology in itself but as his unique contribution to the field of psychology. Jung claimed late in his career that only for about a third of his patients did he use "Jungian analysis."

For another third, Freudian psychology seemed to best suit the patient's needs and for the final third Adlerian analysis was most appropriate. In fact, it seems that most contemporary Jungian clinicians merge a developmentally grounded theory, such as Self psychology or Donald Winnicott's work, with the Jungian theories in order to have a "whole" theoretical repertoire to do actual clinical work.

The "I" or Ego is tremendously important to Jung's clinical work. Jung's theory of etiology of psychopathology could almost be simplified to be stated as a too rigid conscious attitude towards the whole of the psyche. That is, a psychotic episode can be seen from a Jungian perspective as the "rest" of the psyche overwhelming the conscious psyche because the conscious psyche effectively was locking out and repressing the psyche as a whole.

John Weir Perry's book The Farside of Madness explores and fleshes out this idea of Jung's very well. Note: this is a psychological description of a psychotic episode.

Jung hypothesized a medical basis for schizophrenia that was beyond the understanding of the medical science of his day (and seems to still be beyond present medical science in a satisfactory sense). Twin studies and plenty of clinical material seem to point clearly to a medical basis for schizophrenia. It perhaps can best be said that schizophrenia is both medical and psychological. A medical understanding (again, as yet still lacking) would not change the fact that schizophrenia is lived by those who have it psychologically; that is to say, as theorists and scientists, we may be able to say that schizophrenia happens in genes, brains, and the electrochemical, but for one who has schizophrenia it also happens in their mind and experience.

This is to say a purely medical treatment of major mental illness is inadequate, as is a purely psychological treatment of major mental illness.

Post-Jung Theories

Samuels (1985) has distinguished three schools of "post-Jungian" therapy-the classical, the developmental and the archetypal.

Classical School

The classical school is that which tries to remain faithful to what Jung himself proposed and taught in person and in his 20-plus volumes of work.

Developmental School

The developmental school, associated with Michael Fordham, Brian Feldman etc., can be considered a bridge between Jungian psychoanalysis and Melanie Klein's object relations theory. Laings and Goodheart are also often mentioned.

Archetypal School

The archetypal school (sometimes called "the imaginal school"), with different views associated with the Mythopoeticists, such as James Hillman in his intellectual theoretical view of Archetypal psychology, Clarissa Pinkola Estés, in her view that ethnic and aboriginal people are the originators of archetypal psychology and have long carried the maps to the journey of the soul in their songs, tales, dream-telling, art and rituals; Marion Woodman who proposes a feminist viewpoint regarding archetypal psychology, and other Jungians like Thomas Moore, as well. Most mythopoeticists/ archetypal psychology innovators either imagine the Self not to be the main archetype of the collective unconscious as Jung thought, but rather assign each archetype equal value...Others, who are modern progenitors of archetypal psychology (such as Estés), think of the Self as that which contains and yet is suffused by all the other archetypes, each giving life to the other.

Robert L. Moore, one of Jung's most dedicated followers, has explored the archetypal level of the human psyche in a series of five books co-authored with Douglas Gillette, which have played an important role in the men's movement in the United States. R. Moore likes to use computerese, so he likens the archetypal level of the human psyche to the hard wiring of a computer. Our personal experiences of course influence our accessing the archetypal level of the human psyche, but personalized ego consciousness can be likened to the software in a computer (e.g., Microsoft Word).

COUNSELLING PSYCHOLOGY

Counseling psychology as a psychological specialty facilitates personal and interpersonal functioning across the life span with a focus on emotional, social, vocational, educational, health-related, developmental, and organizational concerns. Through the integration of theory, research, and practice, and with a sensitivity to multicultural issues, this specialty encompasses a broad range of practices that help people improve their well-being, alleviate distress and maladjustment, resolve crises, and increase their ability to live more highly functioning lives.

Though closely related to clinical psychology, counseling psychology differs from that field in a several subtle ways. First, counseling psychologists typically focus on less severe psychopathology (e.g., depression and anxiety), while clinical psychologists deal with more seriously disturbed individuals (e.g., those with schizophrenia or personality disorders). Second, counseling psychologists are more likely than clinical psychologists to assume a client-centered or humanistic theoretical approach. Finally, counseling psychology is unique in its attention both to normal developmental issues as well as the problems associated with physical, emotional, and mental disorders. Despite these differences, counseling and clinical psychology are becoming increasingly indistinguishable, leading some to suggest that these fields be combined.

Populations served by counseling psychologists include persons of all ages and cultural backgrounds. Examples of those populations would include late adolescents or adults with career/educational concerns and children or adults facing severe personal difficulties. Counseling psychologists also consult with organizations seeking to enhance their effectiveness or the well-being of their members.

Counseling psychologists adhere to the standards and ethics established by the American Psychological Association and the

American Counseling Association. Counseling psychologists or counselors who work in K-12 school settings also need to follow the ethic standard established by the American School Counselor Association. In Australia there are several associations offering standards and ethics. Each professional or trade association has its preferred qualification and practitioner standards, and also suggesting other requirements.

Methodology

Counseling psychologists participate in a range of activities including teaching, research, psychotherapeutic and counseling practice, career development, assessment, supervision, and consultation. They employ a variety of methods closely tied to theory and research to help individuals, groups and organizations function optimally as well as to mediate dysfunction. Interventions may be either brief or long-term; they are often problem-specific and goal-directed. These activities are guided by a philosophy that values individual differences and diversity and a focus on prevention, development, and adjustment across the life-span which includes vocational concerns.

Employment Settings

Counseling psychologists are employed in a variety of settings depending on the services they provide and the client populations they serve. Some are employed in colleges and universities as teachers, supervisors, researchers, and service providers. Others are employed in independent practice providing counseling, psychotherapy, assessment, and consultation services to individuals, families, groups, and organizations. Additional settings in which counseling psychologists practice include community mental health centers, Veterans Administration Medical Centers and other facilities, family services, health maintenance organizations, rehabilitation agencies, business and industrial organizations and consulting within firms.

Training

Counseling psychologists are trained in graduate programs, generally in APA-accredited doctoral programs. There are about

70 APA accredited counseling psychology doctoral programs. Almost all grant a Ph.D, although a few grant a Psy.D. Most doctoral programs take 5-6 years to complete.

Graduate work in counseling psychology includes coursework in general psychology and statistics, counseling practice, and research. Students must complete an original dissertation at the end of their graduate training. Students must also complete a one year full-time internship at an APA accredited site before earning their doctorate. In order to be licensed to practice, counseling psychologists must meet various requirements set by each state (a declining number of which require postdoctoral supervision), and pass a national licensing exam.

CRITICAL PSYCHOLOGY

Critical psychology is a branch of psychology that is aimed at critiquing mainstream psychology and attempts to apply psychology in more progressive ways, often looking towards social change as a means of preventing and treating psychopathology. One of Critical Psychology's main criticisms of conventional psychology is how it fails to consider or deliberately ignores the way power differences between social classes and groups can impact the mental and physical well-being of individuals or groups of people.

Origins

Critical psychology started in the 1970s in Berlin at Freie Universität Berlin, and the German branch of critical psychology predates and has developed largely separately from the rest of the field.

1960 through 1970

In the 1960s and 1970s the term radical psychology was used by psychologists to denote a branch of the field which rejected conventional psychology's focus on the individual as the basic unit of analysis and sole source of psychopathology. Instead, radical psychologists examined the role of society in causing and treating problems, and looked towards social change as an alternative to therapy to treat mental illness and as a means of preventing psychopathology. Within psychiatry the term anti-psychiatry was often used and now British activists prefer the term critical psychiatry. Critical Psychology is currently the preferred term for the discipline of psychology keen to find alternatives to the way the discipline of psychology reduces human experience to the level of the individual and thereby strips away possibilities for radical social change.

In the 1990's

Starting in the 1990s a new wave of books started to appear on critical psychology, the most influential being the edited book Critical Psychnology by Dennis Fox and Isaac Prilleltensky. Various introductory texts to critical psychology written in the United Kingdom have tended to focus on discourse, but this has been seen by some proponents of critical psychology as a reduction of human experience to language which is as politically dangerous as the way mainstream psychology reduces experience to the individual mind.

Ian Parker

In 1999 Ian Parker published an influential manifesto in both the online journal Radical Psychology and the Annual Review of Critical Psychology. This manifesto argues that Critical Psychology should include the following four components:

1. Systematic examination of how some varieties of psychological action and experience are privileged over others, how dominant accounts of 'psychology' operate ideologically and in the service of power;
2. Study of the ways in which all varieties of psychology are culturally historically constructed, and how alternative varieties of psychology may confirm or resist ideological assumptions in mainstream models;
3. Study of forms of surveillance and self-regulation in everyday life and the ways in which psychological culture operates beyond the boundaries of academic and professional practice; and

4. Exploration of the way everyday 'ordinary psychology' structures academic and professional work in psychology and how everyday activities might provide the basis for resistance to contemporary disciplinary practices.

Critical Psychology Today

There are a now international journals devoted to critical psychology, including the International Journal for Critical Psychology and the Annual Review of Critical Psychology. The journals still tend to be directed to an academic audience, though the Annual Review of Critical Psychology runs as a open-access online journal. There are close links between critical psychologists and critical psychiatrists in Britain through the Asylum Collective. Critical psychology courses and research concentrations are available at Manchester Metropolitan University, Cardiff University, the University of the West of England in Bristol, the University of East London and the University of Adelaide.

CRITICISMS OF CONVENTIONAL PSYCHOLOGY

One of the criticisms of conventional psychology raised by critical psychology is the inattention to power differentials between different groups-examples include between psychiatrists and patients, wealthy groups and the less financially well-off, or industrial lobbyists and the general public. This inattention to power has resulted in conventional psychology tending to assume that how things are is how they should be, that the current state of affairs is the natural state of things. As a result, conventional psychology has a tendency to uphold the status quo, victim-blame and situate problems within individuals rather than the social context they are embedded in.

Extensions

Like many critical applications, critical psychology has expanded beyond Marxist roots to benefit from other critical approaches. Consider ecopsychology and transpersonal psychology. Critical psychology and related work has also sometimes been labelled radical psychology and liberation psychology. In the field of developmental psychology, the work of Erica Burman has been influential. Various sub-disciplines within psychology have begun to establish their own critical orientations. Perhaps the most extensive are Critical Health Psychology and Community Psychology.

International Critical Psychology

Germany

At FU-Berlin, critical psychology was not really seen as a division of psychology and followed its own methodology, trying to reformulate traditional psychology on an unorthodox Marxist base and drawing from Soviet ideas of cultural-historical psychology, particularly Aleksey Leontyev. Some years ago the department of critical psychology at the FU-Berlin was merged into the traditional psychology department.

South Africa

The University of KwaZulu-Natal in Durban, South Africa, is one of few worldwide to offer a masters course in critical psychology. For an overview of critical psychology in South Africa, see Desmond Painter and Martin Terre Blanche's article on Critical Psychology in South Africa: Looking back and looking forwards. They have also now started a critical psychology blog.

United States and Canada

Critical psychology in the United States and Canada has, for the most part, focused on critiques of mainstream psychology's support for an unjust status quo. No departments of critical psychology exist, though critical perspectives are sometimes encountered in traditional universities, perhaps especially within community psychology programs. North American efforts include the 1993 founding of RadPsyNet Radical Psychology Network, the 1997 publication of Critical Psychology: An Introduction (edited by Dennis Fox and Isaac Prilleltensky), and the action-focused PsyACT (Psychologists Acting with Conscience Together).

DISCURSIVE PSYCHOLOGY

Discursive psychology is a school of psychology developed in the 1990s by Jonathan Potter and Derek Edwards at

Loughborough University. It draws on the philosophy of language of the later Wittgenstein, the rhetorical approach of Michael Billig, and the conversation analysis of Harvey Sacks. Discursive psychological studies highlight the way people construct versions of 'mental', 'social' and 'material' events and processes as parts of particular practices.

Discursive psychology started with studies that offered new ways of understanding mainstream topics in social and cognitive psychology such as memory, attribution and attitudes. For example, how are accounts constructed to manage issues of attitude and motive? Edwards and Potter illustrate this with an excerpt from a rape trial.

Note how Counsel and Witness produce competing descriptions of the place, each of which offers inferences to the jury. 'It's where girls and fellas meet' gives an impression of what the clientele are expecting and wanting; while 'people go there' neutralizes that impression. The alternative descriptions imply different attitudes, motives and even moral status. These psychological matters are played out in the competing descriptions offered by the two parties to this interaction. This is what discursive psychology is about.

In the past few years work in discursive psychology has focused on material from real world situations such as relationship counselling, child protection helplines and neighbour disputes. It asks questions such as the following. How does a party in relationship counselling construct the problem as something that the other party needs to work on? Or how does a child protection officer working on a child protection helpline manage the possibly competing tasks of soothing a crying caller and simultaneously elicit evidence sufficient for social services to intervene to help an abused child?

It is philosophically opposed to more traditional cognitivist approaches to language.

ECOLOGICAL PSYCHOLOGY

Ecological psychology (EP) is a term claimed by a number of schools of psychology. However, the two main ones are one on the writings of J. J. Gibson, and another on the work of Roger G. Barker, Herb Wright and associates at the University of Kansas in Lawrence. It should be noted that whereas Gibsonian psychology is always termed Ecological Psychology, the work of Barker (and his followers) is also known as Environmental Psychology. There is a considerable amount of overlap between the two schools, although the Gibsonian approach tends to be more philosophical.

Barker's work was based on his empirical work at the Midwest Field Station. He wrote later: "The Midwest Psychological Field Station was established to facilitate the study of human behavior and its environment in situ by bringing to psychological science the kind of opportunity long available to biologists: easy access to phenomena of the science unaltered by the selection and preparation that occur in laboratories." (Barker, 1968). The study of environmental units (behavior settings) grew out of this research. In his classic work "Ecological Psychology" (1968) he argued that human behaviour was radically situated: in other words, you couldn't make predictions about human behaviour unless you know what situation or context or environment the human in question was in. For example, there are certain behaviours appropriate to being in Church, attending a lecture, working in a factory etc., and the behaviour or people in these environments is more similar than the behaviour of an individual person in different environments. He has since developed these theories in a number of books and articles.

Gibson also stressed the importance of the environment. He argued that animals and humans stand in a 'systems' relation to the environment, such that, to fully explain some behaviour it was necessary to study the environment in which this behaviour took place. The aphorism: "It's not what is inside the head that is important, it's what the head is inside of", is supposed to capture that point. However, especially in his later work, Gibson concentrated more on the nature of cognition itself. He rejected 'indirect' (cognitivist) perception, in favour of 'direct realism'; and also rejected the information processing and cognitivist views of human behaviour. Both schools emphasise 'real world' studies of behaviour as opposed to the artificial environment of the laboratory.

ENVIRONMENTAL PSYCHOLOGY

Environmental psychology is an interdisciplinary field focused on the interplay between humans and their surroundings.

Scope

"Environmental psychology" is arguably the best-known and more comprehensive description of the field.

The field is known by the following names, advanced by different researchers, sometimes used interchangeably, sometimes with recognized gaps and overlaps between the terms: environmental social sciences, architectural psychology, socio-architecture, ecological psychology, ecopsychology, behavioral geography, environment-behavior studies, person-environment studies, environmental sociology, social ecology, and environmental design research. This field draws on work in a number of disciplines including anthropology, geography, ekistics, sociology, psychology, history, political science, planning, architecture, and urban design.

The varied names for the field accurately reflect an ongoing debate about its proper scope, for example, whether or not it includes study of human interaction with the natural environment. "Environmental design" is generally understood to describe design activities focused on sustainability, a different matter. Only a small portion of the built environment is attributable to architects, so a focus on "architectural psychology" is seen as too narrow.

Challenges

Since the late 1940s, the field has seen significant research findings and a fair surge of interest in the late 1970s and early 1980s, but has challenges of nomenclature, obtaining objective and repeatable results, scope, and the fact that some research rests on underlying assumptions about human perception, which is not fully understood.

In the words of Guido Francescato, speaking in 2000, environmental psychology encompasses a "somewhat bewildering array of disparate methodologies, conceptual orientations, and interpretations... making it difficult to delineate, with any degree of precision, just what the field is all about and what might it contribute to the construction of society and the unfolding of history."

Behavior Settings

The first significant findings in environmental psychology can be traced back to researcher Roger Barker, who founded his research station in the tiny Kansas town of Oskaloosa (renamed "Midwest" for publication) in 1947, and ran it for several decades.

From detailed field observations he developed the theory that social settings influence behavior. In a store, people assume their roles as customers; in school and church, proper behavior somehow already resides coded in the place. Barker spent his career expanding on what he called ecological psychology, identifying these behavior settings, and publishing accounts like "One Boy's Day" (1951). Some of the minute-by-minute observations of Kansan children from morning to night, jotted down by young and maternal gradate students, may be the most intimate and poignant documents in social science. The "behavior setting" remains a valid principle which receives serious attention.

Barker argued that the psychologist should use T-Methods (psychologist as 'transducer': i.e. methods which study Man in his 'natural environment') rather than O-Methods (psychologist as "operator" i.e. experimental methods). In other words, he preferred field work and direct observation.

Proxemics

In the mid 1950s anthropologist E. T. Hall wrote "The Hidden Dimension" which developed and popularized the concepts of personal space and his more general name for this field, proxemics. He defined proxemics as, "... the study of how man unconsciously structures microspace-the distance between men in the conduct of daily transactions, the organization of space in his houses and buildings, and ultimately the layout of his towns."

Hall defined and measured four interpersonal "zones":

- intimate (0 to 18 inches)
- personal (18 inches to 4 feet)
- social (4 feet to 12 feet)
- public (12 feet and beyond)

In "The Hidden Dimension" he famously observed that the precise distance we feel 'comfortable' with other being people being near us is culturally determined: Saudis, Norwegians, Milanese and Japanese will have differing notions of 'close'. In one of his best known empirical studies, Hall carried out an analysis of employee reactions to Eero Saarinen's last work, the John Deere World Headquarters Building.

Impact on the Built Environment

Ultimately, environmental psychology is oriented towards influencing the work of design professionals (architects, interior designers, urban planners, etc.) and thereby improving the human environment.

On a civic scale, efforts towards improving pedestrian landscapes have paid off to some extent, involving figures like Jane Jacobs and Copenhagen's Jan Gehl. One prime figure here is the late writer and researcher William H. Whyte and his still-refreshing and perceptive "City", based on his accumulated observations of skilled Manhattan pedestrians, steps, and patterns of use in urban plazas.

No equivalent organized knowledge of environmental psychology has developed out of architecture. Most prominent American architects, led until recently by Philip Johnson who was very strong on this point, view their job as an art form. They see little or no responsibility for the social or functional impact of their designs, which was highlighted with failure of public high-rise housing like Pruitt Igoe.

Environmental psychology has conquered one whole architectural genre, although it's a bitter victory: retail stores, and any other commercial venue where the power to manipulate the mood and behavior of customers, places like stadiums, casinos, malls, and now airports. From Philip Kotler's landmark paper on Atmospherics and Alan Hirsch's "Effects of Ambient Odors on Slot-Machine Usage in a Las Vegas Casino", through the creation and management of the Gruen transfer, retail relies heavily on psychology, original research, focus groups, and direct observation. One of William Whyte's students, Paco Underhill, makes a living as a "shopping anthropologist". Most of this most-advanced research remains a trade secret and proprietary.

LEGAL PSYCHOLOGY

Legal psychology involves the application of empirical psychological research to legal institutions and people who come into contact with the law. Legal psychology is a field which takes basic social and cognitive theories and principles and applies them to issues in the legal system such as eyewitness memory, jury decision-making, investigations and interviewing.

Legal psychology and forensic psychology together form the field more generally recognized as "psychology and law". Following up on earlier efforts by psychologists to address legal issues, psychology and law became a field of study in the 1960s as part of an effort to enhance justice. The American Psychological Association's Division 41, the American Psychology-Law Society, is active, and there are similar societies in Britain and Europe.

There are several legal psychology journals, including Law and Human Behavior, Psychology, Public Policy and Law, Psychology, Crime, and Law, and Journal of Psychiatry, Psychology and Law. In addition, research by legal psychologists is regularly published in more general journals that cover both basic and applied research areas.

Training and Education

Legal psychologists typically hold a Ph.D. in some area of psychology (e.g., clinical psychology, social psychology, cognitive psychology, etc.), and apply their knowledge of that field to the law. Although formal legal training (such as a J.D. or Master of Legal Studies degree) can be beneficial, most legal psychologists hold only the Ph.D. In fact, some argue that specialized legal training dilutes the psychological empiricism of the researcher. For instance, to understand how eyewitness memory "works," a psychologist should be concerned with memory processes as a whole, instead of only the aspects relevant to the law (e.g., lineups, accuracy of testimony).

A growing number of universities offer specialized training in legal psychology as either a standalone Ph.D. program or a joint J.D. / Ph.D. program. A list of American universities that offer graduate training in legal psychology can be found here on the website of the American Psychology-Law Society.

Roles of a Legal Psychologist

Academics and Research

Many legal psychologists work as professors in university psychology departments, criminal justice departments or law schools. Like other professors, legal psychologists generally conduct and publish empirical research, teach various classes, and mentor graduate and undergraduate students. Many legal psychologists also conduct research in a more general area of psychology (e.g., social, clinical, cognitive) with only a tangental legal focus. Those legal psychologists who work in law schools almost always hold a J.D. in addition to a Ph.D.

Expert Witnesses

Psychologists specifically trained in legal issues, as well as those with no formal training, are often called by legal parties to testify as expert witnesses. In criminal trials, an expert witness may be called to testify about eyewitness memory, mistaken identity, competence to stand trial, the propensity of a death-qualified jury to also be "pro-guilt," etc. Psychologists who focus on clinical issues often testify specifically about a defendant's competence, intelligence, etc. More general testimony about perceptual issues (e.g., adequacy of police sirens) may also come up in trial.

Experts, particularly psychology experts, are often accused of being "hired guns" or "stating the obvious." Eyewitness memory experts, such as Elizabeth Loftus, are often discounted by judges and lawyers with no empirical training because their research utilizes undergraduate students and "unrealistic" scenarios. If both sides have psychological witnesses, jurors may have the daunting task of assessing difficult scientific information.

Policy Making and Legislative Guidance

Psychologists employed at public policy centers may attempt to influence legislative policy or may be called upon by state (or national) lawmakers to address some policy issue through empirical research. A psychologist working in public policy might suggest laws or help to evaluate a new legal practice (e.g., the use of sequential lineups v.s. simultaneous lineups).

Trial Consulting

Many psychologists act as trial consultants. No special training nor certification is needed to be a trial consultant, though an advanced degree is generally welcomed by those who would hire the trial consultant. The American Society of Trial Consultants does have a code of ethics for members, but there are no legally binding ethical rules for consultants. Some psychologists who work in academics are hired as trial consultants when their expertise can be useful to a particular case. Other psychologist/consultants work for or with established trial consultant firms. The practice of law firms hiring "in-house" trial consultants is becoming more popular, but these consultants usually can also be used by the firms as practicing attorneys.

Trial consultants perform a variety of services for lawyers, such as picking jurors (usually relying on in-house or published statistical studies) or performing "mock trials" with focus groups. Trial consultants work on all stages of a case from helping to organize testimony, preparing witnesses to testify, picking juries, and even arranging "shadow jurors" to watch the trial unfold and provide input on the trial. There is some debate on whether the work of a trial consultant is protected under attorney-client privilege, especially when the consultant is hired by a party in the case and not by an attorney.

Advisory Roles

Legal psychologists may hold advisory roles in court systems. They may advise legal decision makers, particularly judges, on psychological findings pertaining to issues in a case. The psychologist who acts as a court advisor provides similar input to one acting as an expert witness, but acts out of the domain of an adversarial system.

MEDICAL PSYCHOLOGY

Medical psychology (related to Clinical Health Psychology, Psychosomatic Medicine, and Behavioral Medicine') is a branch of clinical psychology where clinicians have trained in the biological aspects of mental illness in relation to physical illness, and are usually qualified to prescribe medication; this last aspect varies from country to country. It adopts the biopsychosocial approach to medicine, which revolves around the idea that both the body and mind are indivisible, and that disease and illness are not identical. Continuing with this line of thought, all diseases whether of the mind or of the physical body must be treated as if they have both been affected. The intent of Medical Psychology is to apply knowledge from all branches of social, psychological, and biological medicine in the prevention, assessment, and treatment of all forms of physical illness and the adaptation to illness; specific behavioral, psychotherapeutic, and pharmaceutical methods are used to help the person respond to illness and prevent further illness through matching coping and management skills to the person's abilities, character, and personality style.

Background

Some clinical practitioners and philosophers today believe that this Cartesian line of thought is outdated. Some anecdotal evidence has been raised concerning the lack of effective treatment for things such as migraines, pains and cancer and that perhaps the cure should not only be centred on the body but also the mind. Some also raise the argument that most uncurable diseases are brought about not merely by physical ailments, but by mental problems. All evidence is of course unproven and will likely remain as such for a long time to come.

The contribution of medical psychologists to general health care increased with their use as primary care doctors in the world's largest HMO, the California Kaiser Permanente. These psychologists work along with primary care physicians in order to determine optimum treatment plans for all patients with physical and/or mental illnesses.

Education

An important contribution of medical psychology is in the education of patients (psychoeducation) in disease processes. Frequently, such education of the patient and the family insures substantially better adherence to treatment recommendations. Medical psychologists are particularly successful in the treatment of asthma, gastrointestinal illnesses, cardiac conditions, spinal cord and brain injuries, chronic pain, headaches, and addictions (drugs, smoking, eating, alcohol, etc.).

Training in medical psychology entails a doctoral degree (Ph.D., or Psy.D.) in clinical psychology, an internship, and postdoctoral training in one of the branches of medicine traditionally associated with psychological fields: psychosomatic medicine, rehabilitation, neuropsychology (central nervous system functioning-brain functions), substance abuse, pain medicine, among others.

MUSICAL PSYCHOLOGY

Music psychology, or the psychology of music, may be regarded either as a branch of psychology or as a branch of musicology. It aims to explain and understand musical behavior and musical experience. Modern music psychology is mainly empirical: music-psychological knowledge tends to advance primarily on the basis of interpretations of data about musical behavior and experience, which are collected by systematic observation of and interaction with human participants.

The Scope of Music Psychology

Humans spend enormous amounts of time, effort, and money on musical activities. Why? The modern, international field of music psychology is gradually exploring a multitude of issues that surround this central question. Music psychology may be regarded as scientific research about human culture. The results of this research have, and will continue to have, direct implications for matters of general concern: human values, human identity, human nature, and quality of life.

Research Areas

Questions in music psychology are often difficult to answer. It is therefore necessary to subject the research literature to careful quality control procedures. These generally take the form of anonymous expert peer review, which is a standard feature of all leading music-psychological societies, conferences, and journals. Music psychologists investigate all aspects of musical behavior by applying methods and knowledge from all aspects of psychology. Topics of study include for example:

- everyday music listening (while driving, eating, shopping, reading...)
- musical rituals and gatherings (religious, festive, sporting, political...)
- the specific skills and processes involved in learning a musical instrument or singing in a choir
- musical behaviors such as dancing and responding emotionally to music
- development of musical behaviours and abilities throughout the lifespan
- the role of music in forming personal and group identities
- preferences: the reasons why we like some kinds of music and not others
- the structures that we hear within music: melody, phrasing, harmony, tonality, rhythm, meter
- the psychological processes involved in musical performance, including:
  - music reading, including eye movement in music reading
  - improvisation
  - the interpersonal/social aspects of group performance
  - the composition/arrangement of music on paper or with the aid of computers

Related Areas of Psychology

- Biopsychology (including Neuropsychology)
- Perception (including Sensation and Psychoacoustics)

- Cognition (including Language, Thinking, Consciousness, Learning, and memory)
- Motivation and Emotion
- Childhood and Life-span development
- Health (including Stress, Coping, Therapy, and Psychological disorders)
- Personality and Individual Differences
- Skill (including Talent, Creativity, and Intelligence), and
- Social psychology and Cognition.

Relation to Music and Musicology

Music psychology can shed light on non-psychological aspects of musicology and musical practice. For example, music psychology contributes to music theory by investigating the perception of musical structures such as melody, harmony, tonality, rhythm, meter, and form. Research in music history can benefit from psychologically inspired, systematic study of the history of musical syntax, or from psychological analyses of the personalities of composers in relation to the psychological effect of their music. Ethnomusicology can benefit from psychological approaches to the study of music cognition in different cultures. Research is only beginning in many of these promising areas of interaction.

PERFORMANCE PSYCHOLOGY

Performance psychology is the branch of psychology that studies the factors that allow individuals, communities and societies to flourish. Most recently, performance psychology has included the study of the psychological skills and knowledge necessary to facilitate and develop peak performance guidelines into best practice for sport, business, fitness and the performing arts.

The past few years have seen an explosion in the field of performance psychology. This growth has been primarily in the study of performance excellence in sports as applied to the field of business. Important links have been made between world-class, championship individual and team sports performance and business results. In a similar vein, there has been an increase in the interest of coaching top performers and addressing their needs, not just providing remedial coaching for underachievers. During that time, more research has been devoted to understanding the characteristics of high achievers in sports as well as business.

THEORETICAL PSYCHOLOGY

Theoretical psychology is concerned with theoretical and philosophical aspects of the discipline of psychology. It is an interdisciplinary field involving psychologists specialising in, amongst others, cognitive, social, developmental, personality, clinical, perceptual, neurological, biological, evolutionary, historical, political and critical psychology.

In the United States, one group focused on theoretical and philosophical issues in Psychology is The Society for Theoretical and Philosophical Psychology (formerly known as The American Psychological Association, Division for Theory and Philosophy of Psychology or APA Division 24). This society regularly hosts symposia, lectures, roundtable discussion, and poster sessions at the annual APA conference.

TRAFFIC PSYCHOLOGY

Traffic psychology is a young expanding field in psychology. Whereas traffic psychology is primarily related to "the study of the behaviour of road users and the psychological processes underlying that behaviour" (Rothengatter, 1997, 223) as well as to the relation between behaviour and accidents, transportation psychology, sometimes referred to as mobility psychology, has its focus on mobility issues, individual and social factors in the movement of people and goods, and travel demand management (TDM).

There is no single theoretical framework in traffic psychology, but many specific models explaining, e.g., perceptual, attentional, cognitive, social, motivational and emotional determinants of mobility and traffic behaviour. One of the most prominent behavioural models divides the various tasks involved in traffic participation into three hierarchical levels, i.e. the strategic, the tactical and the operational level. The model demonstrates the diversity of decision and control tasks which have to be accomplished when driving a vehicle. However, until now, most of the psychological models have a rather heuristic nature, e.g. risk theories like the risk compensation hypothesis, Fuller's task capability model, and thus are not sufficiently precise to allow for concrete behavioural prediction and control.

This is partly due to the importance of individual differences, a major topic of psychology which in traffic and transportation has not yet been sufficiently accounted for. On the other hand, social-psychological attitude-behaviour models, such as Ajzen's theory of planned behavior, have been helpful in identifying determinants of mobility decisions.

Bringing together the scientific and practical lines, six areas of traffic and transportation psychology can be distinguished (Schlag, 1999):

Areas of Traffic Psychology

Behaviour and Accident Research

Particularly in relation to different groups of road users (age groups, modes of transport), but also in relation to road design and motor vehicles. Explaining and predicting road user behaviour depends on the development of valid and reliable models about the role of human factors in mobility behaviour and especially driver performance. Psychological traffic accident and behaviour research deals with, e.g.,

- analysis of the driving task, changing conceptually from a traditionally rather sensomotoric task to a task with high monitoring impact,
- perception, cognition and attentiveness when driving, driver information processing and expectations,
- driver state, workload, alertness and fatigue,
- driver personality, risk-taking, attitudes, motives for driving, arousal and emotion,
- interactions and the social psychology of driving,

- the relation between the personal and environmental background of behaviour, overt behaviour, emerging conflicts and accidents,
- work on risk compensation theory.

Accident Prevention and Improvement of Traffic Safety

Education and information, but overall following the "4 E's": enforcement, education, engineering, encouragement/economy. Main goal is promoting safety by influencing and modifying behaviour with legal, educational, vehicle-and road-specific measures; driver training, driving-instructor education, information on traffic issues, campaign design and marketing, effective enforcement.

Mobility, Transport Economy and Engineering

Main objective is user-oriented and best-usable supply and design. This includes differentiation between transportation needs of special groups (elderly, handicapped, young people etc.). Main topics are:

- mobility needs and travel demands, choice of means of transport,
- travel behaviour research, above all activity-based approaches,
- altering mobility behaviour and modal split, problems of habituation and resistance to change, car dependence,
- design and acceptance of travel demand management, above all of road pricing measures (Schade & Schlag, 2003),
- psychological aspects in road design and traffic environment,
- quality management, especially quality of service, usability and well-being.

Vehicle Construction and Design

Psychology in car manufacturing traditionally deals with questions of ergonomics, but since the 1980´s new in-car devices as well as related new infrastructure has emerged as a rapidly growing field. Advanced Driver Assistance Systems (ADAS) and new information systems are designed to support the driver in a suitable and user-oriented way. Based on analyses of driving tasks which drivers have to cope with, e.g. multiple tasks requiring divided attention, psychologists' primary orientation in the design process is towards human needs defining the technical requirements, human-centred development, usability of ADAS, operability of Human-Machine Interfaces (HMI), behavioural adaptation and risk compensation, acceptance of innovations, and social impacts.

Psychological assessment and counselling/rehabilitation for drivers who have become conspicuous: driver assessment, training and rehabilitation, above all for drivers with offences (driving while intoxicated, severe offences against traffic laws), aptitude assessment for driving, selection and training for professional drivers.

Rail and Flight Psychology

Parts of the mentioned domains not only apply to road traffic but also to rail and air transport. Nevertheless, rail and flight psychology have historically developed in part separately from the dominantly road-related traffic psychology. One major new direction in rail as well as in flight psychology is the focus shift from the professional operator (selection and training) to the customer perspective (quality of service, usability).

From its very beginning, traffic psychology in research and practice has followed an interdisciplinary approach and has shared common topics especially with medicine (e.g. related to driving aptitude), engineering (ergonomics of cars as well as human factors in traffic planning), and economics (e.g. travel demand management). People as traffic participants are seen as the core of an interactive traffic system also comprising transportation means, routes, traffic environment and regulation. Thus mobility, including its positive as well as detrimental impacts, has its origin in people's desires, decisions and behaviour-and these might be influenced. The main accident causes are human errors and maladaptive behaviour, accounting alone or in interaction with roadway or vehicle-related causes for more than 90% of all traffic accidents. Recognizing the possible impact of psychology in studying and solving transport problems, traffic and transportation psychology has emerged rapidly since the 1980's.

CULTURAL-HISTORICAL PSYCHOLOGY

Cultural-historical psychology (the school of Vygotsky)-a trend in psychological research founded by Lev Vygotsky in the end of the 1920s and developed by his students and followers in Eastern Europe and worldwide. Cultural-historical psychology emerged as a response to Cartesian dualism between mind and body in psychology of that time as a deliberate attempt to establish a new paradigm in psychological resarch that would overcome the narrow objectivism of behaviourism and subjectivism of introspective psychology of Wundt, James, and others.

Vygotsky and his associates postulate in principle non-adaptive character and the mechanisms of higher psychical (mental) functions development. Defining the main goal of psychological inquiry as an objective study of human consciousness, the members of Vygotsky's school investigate the role of cultural mediation and such cultural mediators as word, sign (Vygotsky), symbol, myth (Losev, V. Zinchenko) in the development of human higher psychical functions, development of personality and its "top-most' phenomenology.

Some of Vygotsky's students who took up and developed his approach to psychology include Aleksey Leontyev (sometimes also spelled A.N. Leontev) and A.R. Luria in the Soviet Union, and Klaus Holzkamp in Germany. Holzkamp developed his own approach to cultural-historical psychology which he termed "critical psychology" mainly based on Leontev's work, also making use of Merleau-Ponty's Phenomenology.

EGO PSYCHOLOGY

Ego psychology is a school of psychoanalysis that originated in Sigmund Freud's ego-id-superego model.

History

After Freud, a number of prominent psychoanalytic theorists began to elaborate on Freud's functionalist version of the ego.

They put much effort into theorizing the ego's various functions and how they can be impaired in psychopathology. Much of their work focused around strengthening the ego so it could better cope with the pressures from the id, super-ego, and society in general.

The central functions of the ego were traditionally seen as reality-testing, impulse-control, judgment, affect tolerance, defence, and synthetic functioning. An important conceptual revision to Freud's structural theory was made when Heinz Hartmann argued that the healthy ego includes a sphere of autonomous ego functions that are independent of mental conflict. Memory, motor coordination, and reality-testing, for example, ought to be able to function without the intrusion of emotional conflict. According to Hartmann, psychoanalytic treatment aims to expand the conflict-free sphere of ego functioning. By doing so, Hartmann believed, psychoanalysis facilitates adaptation, that is, more effective mutual regulation of ego and environment.

David Rapaport systematized Freud's structural model and Hartmann's revisions of it. Rapaport argued that the central principle of Freudian theory is that mental processes are motivated and shaped by the need to discharge tension. Clarifying Freud's work, Rapaport portrayed the mind as divided into drives and structures. Drives contain fluid energy that pushes for rapid discharge through the immediate gratification of wishes. Because it is rare that wishes can actually be immediately gratified, the mind develops the capacity to delay gratification or achieve it through detours. Consequently, drive energy becomes tied up in the relatively stable mental structures comprising the ego. Rapaport defined structures as mental organizations with a slow rate of change, slow in comparison with the more fluid drives.

Arlow and Brenner argued that Freud's earlier theory of the conscious, preconscious, and unconscious systems of the mind ought to be abandoned, and the structural model used as the sole psychoanalytic theory of the mind.

Recent ego psychological authors have taken the approach in a number of directions. Some, such as Charles Brenner, have contended that the structural model should be abandoned and psychoanalysts should focus exclusively on understanding and treating mental conflict. Others, such as Frederic Busch, have argued for an increasingly nuanced and sophisticated concept of the ego.

Ego psychology is often confused with self psychology, which emphasizes the strength and cohesion of a person's sense of self. Although some ego psychologists write about the self, they usually distinguish the self from the ego. They define the ego as an agency comprised of mental functions, whereas the self is an internal representation of how one sees oneself. In ego psychology, emphasis is placed on understanding the functioning of the ego and its conflicting relations to the id, superego, and reality, rather than on the subjective sense of self.

Defense Analysis

The clinical technique most commonly associated with ego psychology is defense analysis. Through clarifying, confronting, and interpreting the typical defense mechanisms a patient uses, ego psychologists hope to help the patient gain control over these mechanisms.

Criticisms of Ego Psychology

Many authors have criticized Hartmann's conception of a conflict-free sphere of ego functioning as both incoherent and inconsistent with Freud's vision of psychoanalysis as a science of mental conflict. Freud believed that the ego itself takes shape as a result of the conflict between the id and the external world. The ego, therefore, is inherently a conflicting formation in the mind. To state, as Hartmann did, that the ego contains a conflict-free sphere may not be consistent with key propositions of Freud's structural theory.

Some have also accused Hartmann of proposing a conformist psychology in which the ego is considered most healthy when it adjusts to the status quo. Hartmann claimed, however, that his aim was to understand the mutual regulation of the ego and environment rather than to promote adjustment of the ego to the environment.

Also, Jacques Lacan, a prominent psychoanalyst, had a certain disdain for ego-psychology. He took issue with the movement insofar as his form of psychoanalysis focuses on the unconscious. It also splits the ego and theorizes how one never has a true relation to their ego because it is an illusionary relationship to an ideal image, and is a product of the unconscious itself.

GESTALT PSYCHOLOGY

Gestalt psychology (also Gestalt theory of the Berlin School) is a theory of mind and brain that proposes that the operational principle of the brain is holistic, parallel, and analog, with self-organizing tendencies. The classic Gestalt example is a soap bubble, whose spherical shape (its Gestalt) is not defined by a rigid template, or a mathematical formula, but rather it emerges spontaneously by the parallel action of surface tension acting at all points in the surface simultaneously. This is in contrast to the "atomistic" principle of operation of the digital computer, where every computation is broken down into a sequence of simple steps, each of which is computed independently of the problem as a whole. The Gestalt effect refers to the form-forming capability of our senses, particularly with respect to the visual recognition of figures and whole forms instead of just a collection of simple lines and curves.

Properties

The key properties of Gestalt systems are emergence, reification, multistability, and invariance.

Emergence

Emergence is demonstrated by the perception of the Dog Picture, which depicts a Dalmatian dog sniffing the ground in the shade of overhanging trees. The dog is not recognized by first identifying its parts (feet, ears, nose, tail, etc.), and then inferring the dog from those component parts. Instead, the dog is perceived as a whole, all at once.

Reification

Reification is the constructive or generative aspect of perception, by which the experienced percept contains more explicit spatial information than the sensory stimulus on which it is based. For instance, a triangle will be perceived in picture A, although no triangle has actually been drawn. In pictures B and D the eye will recognise disparate shapes as "belonging" to a single shape, in C a complete three-dimensional shape is seen, where in actuality no such thing is drawn.

Invariance

Invariance is the property of perception whereby simple geometrical objects are recognized independent of rotation, translation, and scale; as well as several other variations such as elastic deformations, different lighting, and different component features. For example, the objects in A in the figure are all immediately recognized as the same basic shape, which are immediately distinguishable from the forms in B. They are even recognized despite perspective and elastic deformations as in C, and when depicted using different graphic elements as in D.

Web-based forums and email providers rely on invariance of human perception to prevent automated bots from exploiting the services. A CAPTCHA test presents a distorted image of letters and numbers, not readable by computers, and prompts user to correctly type the string.

Emergence, reification, multistability, and invariance are not separable modules to be modeled individually, but they are different aspects of a single unified dynamic mechanism.

For a mathematical example of such a mechanism using the cubes of psychologists' block design tests, see Block Designs in Art and Mathematics and The Kaleidoscope Puzzle.

Origins

Although Max Wertheimer is credited as the founder of the movement, the concept of Gestalt was first introduced in contemporary philosophy and psychology by Christian von Ehrenfels (a member of the School of Brentano). The idea of Gestalt has its roots in theories by Johann Wolfgang von Goethe, Immanuel Kant, and Ernst Mach. Wertheimer's unique contribution was to insist that the Gestalt is perceptually primary, defining the parts of which it was composed, rather than being an "additional" element over and above the components parts, as von Ehrenfels earlier Gestalt-qualität had been.

Both von Ehrenfels and Edmund Husserl seem to have been inspired by Mach's work Beiträge zur Analyse der Empfindungen (Contributions to the Analysis of the Sensations, 1886), in formulating their very similar concepts of Gestalt and Figural Moment, respectively.

Early 20th century theorists, such as Kurt Koffka, Max Wertheimer, and Wolfgang Köhler (students of Carl Stumpf) saw objects as perceived within an environment according to all of their elements taken together as a global construct. This 'gestalt' or 'whole form' approach sought to define principles of perception--seemingly innate mental laws which determined the way in which objects were perceived.

These laws took several forms, such as the grouping of similar, or proximate, objects together, within this global process. Although it has been criticized for being merely descriptive, it has formed the basis of much further research into the perception of patterns and objects (ref: Carlson, Buskist & Martin, 2000), and of research into behavior, thinking, problem solving and psychopathology.

Theoretical Framework and Methodology

The investigations developed at the beginning of the 20th century, based on traditional scientific methodology, divided the object of study into a set of elements that could be analyzed separately with the objective of reducing the complexity of this object. Contrary to this methodology, the school of Gestalt practiced a series of theoretical and methodological principles that attempted to redefine the approach to psychological research.

The theoretical principles are the following:

- *Principle of Totality:* The conscious experience must be considered globally (by taking into account all the physical and mental aspects of the individual simultaneously) because the nature of the mind demands that each component be considered as part of a system of dynamic relationships.

- *Principle of Psychophysical Isomorphism:* A correlation exists between conscious experience and cerebral activity.

Based on the principles above the following methodological principles are defined:

- *Phenomenon Experimental Analysis:* In relation to the Totality Principle any psychological research should take as a starting point phenomena and not be solely focused on sensory qualities.
- *Biotic Experiment:* The School of Gestalt established a need to conduct real experiments which sharply contrasted with and opposed classic laboratory experiments. This signified experimenting in natural situations, developed in real conditions, in which it would be possible to reproduce, with higher fidelity, what would be habitual for a subject.

Pragnanz

The most basic rule of gestalt is the law of prägnanz. This law says that we try to experience things in as good a gestalt way as possible. In this sense, "good" can mean several things, such as regular, orderly, simplistic, symmetrical, etc. The other gestalt laws are:

- *Law of Closure:* Our mind adds missing elements to complete a figure.
- *Law of Similarity:* Our mind groups similar elements to an entity. The similarity depends on relationships constructed about form, color, size and brightness of the elements.

Law of Similarity

- *Law of Proximity:* Regional or chronological closeness of elements are grouped by our mind and seen as belonging together.
- *Law of Symmetry:* Symmetrical images are seen as belonging together regardless of their distance.
- *Law of Continuity:* The mind continues a pattern, even after it stops.

- *Law of Common Fate:* Elements with the same moving direction are seen as a unit.

Law of Proximity

Figure-ground minds have an innate tendency to perceive one aspect of an event as the figure or foreground and the other as the ground or the background.

Under the gestalt theory, these laws not only apply to images, but to thought processes, memories, and our understanding of time.

Examples of the Gestalt experience include the perception of an incomplete circle as a whole or a pattern of dots as a shape-the mind completes the missing pieces through extrapolation. Studies also indicate that simple elements/ compositions where the meaning is directly perceived do not offer as much a challenge to the mind as complex ones and hence the latter are preferred over the former.

Gestalt Views in Psychology

Gestalt psychologists find it is important to think of problems as a whole. Max Wertheimer considered thinking to happen in two ways: productive and reproductive.

Productive thinking-is solving a problem with insight.

Reproductive thinking-is solving a problem with previous experiences and what is already known. (1945/1959)

Other Gestalts psychologist Perkins believes insight deals with three processes:

(1) Unconscious leap in thinking..

(2) The increased amount of speed in mental processing.

(3) The amount of short-circuiting which occurs in normal reasoning.

Other views going against the Gestalts psychology are:

(1) Nothing-Special View

(2) Neo-Gestalts View

(3) The Three-Process View

Relationship to Gestalt Therapy

Gestalt psychology should not be confused with the Gestalt therapy of Fritz Perls, which is only peripherally linked to Gestalt psychology. A strictly Gestalt psychology-based therapeutic method is Gestalt Theoretical Psychotherapy, developed by the German Gestalt psychologist and psychotherapist Hans-Juergen Walter.

Applications in Computer Science

The gestalt laws are used in user interface design. The law of similarity and law of proximity can for example be used as guides for placing radio buttons. Gestalt psychology also has applications in computer vision for trying to make computers "see" the same things as humans do.

11

Psychological Adjustment

Adjustment (from late Latin ad-juxtare, derived from juxta, near, but early confounded with a supposed derivation from Justus, right), regulating, adapting or settling; in commercial law, the settlement of a loss incurred at sea on insured goods.

All human behaviour is purposeful. These are directed towards the attainment of some goals or satisfaction of some need. An individual is said to be adjusted in environment if there is harmony among his needs are attainable. The other ting is that these goals should be socially desirable. So it is the harmony of the internal with the external.

"Life presents a continuous chain of struggle for existence and survivals" says Darwin. Life is a continuous process of overcoming difficulties or of making adjustments.

General meaning of adjustment is the process of living itself or a dynamic equilibrium of the total personality. It is a life-long process in which one enters into a relationship of harmony with one's environment. Psychologically, adjustment means a person interacts with his/her environment. Fortunate is the individual who is adjusted and considers it so. Every individual, great or small, old or young, is confronted with the problem of adjustment. The problem of adjustment starts right from the birth of the child and continues till his death. The problem of adjustment is both internal as well as external. The problem of adjustment is related to arriving at a balanced state between he needs of the individual and their satisfaction. Needs of the individual are multidimensional. Adjustment is a relative term. Opposite of adjustment is maladjustment. Life presents a continuous chain of struggle for adjustment.

MEANING AND DEFINITION

Dictionary word meaning-Make suitable, adopt.

- James. Drever (1952): Adjustment means the modification to compensate for or meet special conditions.
- Carter V. Good (1959): Adjustment is the process of finding and adopting modes of behaviour suitable to the environment or the changes in the environment.
- Warren (1934): Adjustment refers to any operation whereby an organism or organ becomes more favourably related to the environment or to the entire situation, environmental and internal.
- Gates and Jersild (1948): Adjustment is a continual process in which a person varies his behaviour to produce a more harmonious relationship between himself and his environment.
- Crow and Crow (1956): As individual adjustment is adequate wholesome or healthful to the extent that he has established harmonious relationship between himself and the conditions, situations and persons who comprise his physical and social environment.
- L. S. Shaffer: Adjustment is the proves by which living organism maintains a balance b/w its need and the circumstances that influences the satisfaction of these needs.
- Vonhaller: We can think of adjustment as psychological survival in which the same way as biologist uses the term adaptation to describe physiological.
- Boring: Adjustment as a process by which a living org maintained balance between its needs and the circumstances that influences the satisfactions of these needs.
- According to Laurance F. Shaffer, "Adjustment is the process by which a living organism maintains a balance between its needs and the circumstances that influences the satisfaction of these needs."
- According to Coleman, James C., "Adjustment is the outcome of the individual's attempts to deal with stress

and meet his needs: also, his efforts to maintain harmonious relationship with the environment."

- H. C. Smith: "A good adjustment is one which is both realistic and satisfying. At least in the long run, it reduces to a minimum the frustrations, the tension and anxieties which a person must endure."
- Traxler, "Occasionally in the use of the term adjustment we imply that the most desirable state of adjustment is one in which the individual is perfectly happy and satisfied with all aspects of his life and one in which he has reached the level in all his contacts with his contacts with his environment that he would be glad to see persist thorough his life."

A perusal of the above definitions of adjustment leads us to the following characteristics of adjustment:

1. It helps us to keep balance between our needs and the capacity to meet these needs.
2. It implies changes in our thinking and way of life to the demands of the situation.
3. It gives us the ability and strength to bring desirable changes in the state of our environment.
4. It is physiological as well as psychological.
5. It is multidimensional.
6. It brings us happiness and contentment.

Therefore, a comprehensive definition of adjustment would be like this: "Adjustment is a condition or state in which one feels that one's needs have been (or will be) fulfilled and one's behaviour conforms to the needs of a given environment or the environment is changed (or will be changed) in a manner as it conforms to the needs of the individual.

Learning Exercise 1.

Note:

a) Give your answer as instructed in each question.

b) Compare your answer with above sub section.

Q.1. Define Adjustment.

Ans. ....

....

Q.2. List the main characteristics of adjustment.

Ans. ....

....

Q.3. What is the dictionary meaning of adjustment ?

Ans. ....

....

Q.4. Define adjustment in words of Crow and Crow (1956).

Ans. ....

....

CONCEPT OF ADJUSTMENT

The concept of adjustment means adaptation to physical environment as well as to social demands. No human being can live apart from his physical environment. There is action and reaction chain going on between the individual and his environment. Then there are social pressures and demands of socialization. To these may be added the individual's personal demand such as the satisfaction of physiological needs. All this complex functioning of the persons demands adjustment. The process of adjustment becomes still more complicated when his interaction with one situation comes into conflict with the requirements of the other situation. One situation may give rise to pleasure while the other may give rise to pain. The resulting tension may cause disturbance in his psyche, produce uncomfortable physical symptoms or may even lead to abnormal behaviour. The concept of adjustment is as old as human race on earth. Systematic emergence of this concept starts from Darwin. In those days the concept was purely biological and he used the term adaptation. The adaptability to environmental hazards goes on increasing as we proceed on the phologenetic scale from the lower extreme to the higher extreme of life. Insects and germs, in comparison to human beings, cannot withstand the hazards of changing conditions in the environment and as the season changes, they die. Hundreds of species of insects and germs perish as soon as the winter begins. Man, among the living beings, has the highest capacities to adapt to new situations. Man as a social animal not only adapts to physical demands but he also adjusts to social pressures in the society. Biologists used the term adaptation strictly for physical demands of the environment but psychologists use the term adjustment for varying conditions of social or inter personal relations in the society. Thus, we see that adjustment means reaction to the demands and pressures of social environment imposed upon the individual has to react. Observe the life of a child, he is asked to do this and not to do other things. He has to follow certain beliefs and set of values which the family follows. His personality develops in the continuous process of interaction with his family environment. There are other demands which may be termed as internal as hunger, water, oxygen and sleep etc. If we do not fulfill these internal demands, we feel uncomfortable. With the development of the child, these physiological demands go on increasing and become more complex. These two types of demands sometimes come into conflict with each other and resultantly make the adjustment a complicated process for the individual. Conflicts among the various needs or demands of a person present special problems of adjustment. If you gratify one of the conflicting needs, the need which is not gratified will produce frustration and leads sometimes to abnormal behaviour. Psychologists have interpreted adjustment from two important point of views. One, adjustment as an achievement and another, adjustment as a process. The first point of view emphasizes the quality or efficiency of adjustment and the second lays emphasis on the process by which an individual adjusts in his external environment. Now, let us examine both of these approaches in detail.

Learning Exercise 2.

Note:

a) Give your answer as instructed in each question.

b) Compare your answer with above sub section.

Q.1. Writ a note on (a) adaptation, (b) Darwin's concept of Adjustment, (c) Nature of Adjustment.

Ans. ....

....

....

Q.2. Describe Concept of Adjustment as interpreted by biologists.

Ans. ....

....

Nature of Adjustment

The concept of adjustment was originally biological one and was concerned with adaptation to physical environment for survival. Adaptation to physical environment is, of course, a person's important concern, but he has also to adjust to social pressures and demands of socialization that are inherent in living interdependently with other persons. There are also the demands from a person's internal nature, his physiological needs like hunger, thirst, sleep, sex, elimination, etc. and psychological needs like needs to belong to get esteem, to self actualize, to get in combination and in interactive fashion that influence the psychological functioning and adjustment of person.

Process of Adjustment

The process of adjustment is complicated because a person's interaction with one demand may come in conflict with the requirement of another. Conflict can arise either because two internal needs are in opposition, or because two external demands are incompatible with each other, or because an internal need opposes an external demand. Conflict presents special problems of adjustment. Satisfaction of one need as opposed to other needs may not provide full satisfaction. On the other hand, failure to gratify a strong need or to respond to a strong external demand may result in painful tensions. These tensions can disturb psychological comfort, produce physical symptoms, or result in abnormal behaviour. (a)

Adjustment as an Achievement and as a Process: Adjustment may be viewed from two angles. From one angle, adjustment may be viewed as achievement or how well a person handles his conflicts and overcomes the resulting tension. From another angle, adjustment may be looked upon as a process or how a person adjusts to his conflicts. In the first case we ask whether a person's adjustment is adequate and efficient ? In the second case we ask how does he adjust or what are the modes of adjustment to various demands. (b) Adjustment as a Psychological Process: Adjustment as a process is of major interest to psychologists who want to understand a person and his behaviour. The way one tried to adjust himself and to his external environment at any point of time depends upon interaction between the biological factors in growth and his social experiences. In general, there are three broad types of adjustive process in the event of a conflict between a person's internal need states and environment demands: (a) The person may modify or inhibit the internal impulse. (b) The person may try to alter the environmental demand in some manner so that he resolves the conflict. (c) The person may "escape" through unconscious resources to mental mechanisms like phantasy, compensation, projection, rationalization, sublimation, etc. We cannot call any of these modes of adjustments as the most superior. None of them used in isolation, to the exclusion of others is helpful in adjustment. Excessive use of any one of them is likely to be maladoptive. The human beings in order to reconcile their needs or the environmental demands must modify or inhibit their own impulses sometimes, alter or modify the environment at other times, and use some mental mechanism at other times and at times a combination of all the three.

Learning Exercise 3.

Note:

a) Give your answer as instructed in each question.

b) Compare your answer with above sub section.

Q.1. Explain adjustment as a psychological process.

Ans. ....

....

Principles of Adjustment

The principles of adjustment are being explained here:

(a) Principles of knowing the self-To be adjusted, one should be aware of his strengths and weakness so that he may mould himself according to the required life style.

(b) Principle of accepting one's self-An individual should accept himself as he is. If he is not satisfy with himself and does not respect himself, he cannot adjust in the environment. He always complains of his bad luck and does not have confidence in his capabilities.

(c) Principle of integrating the self-The personality of an individual should be integrated. Those who have dis-integrating personality and are not able to take decision and do not stick to a particular decision, they can not adjust easily.

(d) Principle of self drive and shaping-Sometimes the decisions are imposed on individual. They not opt for a particular career or a business. These decisions are imposed by some influential person. They feel disturbed and maladjusted.

(e) Principle of self control-The aggressive control and imposed discipline is the reason of maladjustment in an individual. So we should advocate self discipline.

(f) Principle of balance and harmonious development-An individual should aim at physical, mental, social, emotional and moral development. Then he may be able to adjust himself.

(g) Principle of understanding others-An individual should not only try to understand himself but also the others. He should respect their ideas and emotions. (Dr. Kuljeet Brar, Chandigarh)

CLOSURE (PSYCHOLOGY)

In psychology, closure may refer to the state of experiencing an emotional conclusion to a difficult life event, such as the breakdown of a close interpersonal relationship or the death of loved one. People may behave in a certain way or perform certain rituals to help "bring closure" following such events. In many other cases, "closure" may only come about through the passage of time (as in "time heals all wounds").

The term is also used in social construction of technology (SCOT) as described by Wiebe Bijker, Trevor Pinch and John Law, among others. The focus of SCOT-studies is to examine the ways in which technological artifacts, which are part of broader technological systems, are socially constructed by so-called relevant social groups or actors. This in essence often means explaining why an artifact looks the way it does. Closure here refers to the end of the process of social construction, meaning that the artifact no longer goes through dramatic changes. Bijker has exemplified this with the development of the bicycle.

CHARACTER STRUCTURE

A character structure is a system of relatively permanent motivational and other traits that are manifested in the characteristic ways that an individual relates to others and reacts to various kinds of challenges. The word "structure" indicates that these several characteristics and/or learned patterns of behavior are linked in such a way as to produce a state that can be highly resistant to change. The idea has its roots in the work of Sigmund Freud and several of his followers, the most important of whom (in this respect) are Wilhelm Reich and Erich Fromm. Among other important participants in the establishment of this concept must surely be counted Erik Erikson.

Among the earliest factors that determine an individual's eventual character structure are his or her genetic characteristics and early childhood nurture and education. A child who is well nurtured and taught in a relatively benign and consistent environment by loving adults who intend that the child should learn how to make objective appraisals regarding the environment will be likely to form a normal or productive character structure. On the other hand, a child whose nurture and/or education are not ideal, living in a treacherous environment and interacting with adults who do not take the long-term interests of the child to heart will be more likely to form a pattern of behavior that suits the child to avoid the challenges put forth by a malign social environment. The means that the child invents to make the best of a hostile environment. Although this may serve the child well while in that bad environment, it may also cause the child to react in inappropriate ways, ways damaging to his or her own interests, when interacting with people in a more ideal social context. Major trauma that occurs later in life, even in adulthood, can sometimes have a profound effect. However, character may also develop in a positive way according to how the individual meets the psychosocial challenges of the life cycle (Erikson).

Freud's first paper on character described the anal character consisting of stubbornness, stinginess and extreme neatness. He saw this as a reaction formation to the child's having to give up pleasure in anal eroticism.The positive version of this character is the conscientious, inner directed obsessive. Freud also described the erotic character as both loving and dependent. And the narcissistic character as the natural leader, aggressive and independent because of not internalizing a strong super ego.

For Erich Fromm character develops as the way in which an individual structures modes of assimilation and relatedness.The character types are almost identical to Freud's but Fromm gives them different names, receptive, hoarding, exploitative. Fromm adds the marketing type as the person who continually adapts the self to succeed in the new service economy. For Fromm, character types can be productive or unproductive. Fromm notes that character structures develop in each individual to enable him or her to interact successfully within a given society, to adapt to its mode of production and social norms may be very counter-productive when used in a different society.

DAYDREAM

A daydream is a fantasy that a person has while awake, often about spontaneous and fanciful thoughts not connected to the person's immediate situation. There are so many different types of daydreaming that there is still no consensus definition amongst psychologists. While daydreams may include fantasies about future scenarios or plans, reminiscences about past experiences, or vivid dream-like images, they are often connected with some type of emotion.

Daydreaming may take the form of a train of thought, leading the daydreamer away from being aware of his immediate surroundings, and concentrating more and more on these new directions of thought. To an observer, they may appear to be affecting a blank stare into the distance, and only a sudden stimulus will startle the daydreamer out of their reverie.

While daydreaming has long been derided as a lazy, non-productive pastime, as can be seen in the use of the derogatory phrase "pipe dream," daydreaming can be constructive in some contexts. There are numerous examples of people in creative or artistic careers, such as composers, novelists, and filmmakers, developing new ideas through daydreaming. Similarly, research scientists, mathematicians, and physicists have developed new ideas by daydreaming about their subject areas.

Of course, excessive daydreaming may be bad for some individuals, such as severely depressed people who daydream about their shortcomings, thus miring themselves further in their depression. However, people who daydream more than average may have some psychological strengths, such as increased empathy. Some psychologists use the mental imagery created during their clients' daydreaming to help gain insight into their mental state.

Historical Backdrop

Daydreaming was long held in disrepute in the Western world and was associated with laziness. In the late 1800s, Sigmund Freud argued that some daydreams with grandiose fantasies are self-gratifying attempts at "wish fulfillment.In the 1950s, some educational psychologists warned parents not to let their children daydream, for fear that the children may be sucked into "neurosis and even psychosis."

In the late 1960s, psychologist Jerome L. Singer of Yale University and psychologist John S. Antrobus of the City College of New York created a daydream questionnaire. The questionnaire, called the Imaginal Processes Inventory (IPI) has been used to investigate daydreams. Psychologists Leonard Giambra and George Huba used the IPI and found that daydreamers' imaginary images vary in three ways: how vivid or enjoyable the daydreams are, how many guilt-or fear-filled daydreams they have, and how "deeply" into the daydream people go.

Recent Research

Eric Klinger's research in the 1980s showed that most daydreams are about ordinary, everyday events and help to remind us of mundane tasks. Klinger's research also showed that over 3/4 of workers in boring jobs, such as lifeguards and truck drivers, use vivid daydreams to "ease the boredom" of their routine tasks. Klinger found that less than five percent of the workers' daydreams involved explicitly sexual thoughts and that violent daydreams were also uncommon.

Israeli high school students who scored high on the daydreaming scale of the IPI had more empathy than students who scored low. Some psychologists, such as Los Angeles' Joseph E. Shorr, use the mental imagery created during their clients' daydreaming to help gain insight into their mental state and make diagnoses.

EMOTIONAL COMPETENCE

Emotional competence refers to a person's competence in expressing or releasing their emotions. It implies an ease around emotions which results in emotionally competent people being relaxed about other people being emotional.

Description

The concept of emotional competence is rooted in the understanding of emotions as being normal, useful aspects of being human. Anger is a reaction to aggression and gives a person the strength to repel the aggression. Grief is a reaction to abandonment or feeling unloved and it has the effect of eliciting sympathetic responses from others. Fear is a response to danger and has a clear physiological effect of heightening our senses and speeding up our reactions.

From this it can be seen that the suppression of emotion is not useful and that teaching people to suppress their emotions is part of trying to control them. Emotionally competent people will express emotion appropriate to the situation and their needs and they will not seek to suppress emotions in others.

It is fairly widely believed that if appropriate emotions are not expressed some sort of memory of them becomes stored. Later events may trigger off the old emotions resulting in inappropriate emotional responses. This particularly applies to emotions that children are prevented from expressing. Releasing these old emotions is a key feature of co-counselling.

Emotional competence can lead to improved health through avoiding stress that would otherwise result from suppressing emotions. It can also lead to improved relationships since inappropriate emotions are less likely to be expressed and appropriate behaviour is not avoided through fear of triggering some emotion.

The concept is distinct from emotional intelligence which, while recognising the importance of emotions, gives emphasis to controlling or manipulating them.

Assertiveness and Emotional Competence

Humanistic approaches to assertiveness, as for instance outlined by Anne Dickson (Dickson, 1982) emphasise the importance of working with emotions. In particular it recognises the need to address manipulative or passive (the person does not say what they want)-aggressive (they try to force the other person to do what they want) behaviour in which the manipulator exploits the feelings of the other to try to get what they want. Building up emotional competence is a way of learning to handle such behaviour.

Another aspect is learning to be assertive when feeling emotional. Assertiveness training involves learning a range of ways to handle any situation so that a person is able to choose a way which seems appropriate for them on each occasion. With respect to emotions, people are encouraged to notice and accept what they feel. They then have choices from handling the situation calmly through doing so and saying how they feel to letting the emotion out, all of which involve emotional competence.

FANTASY (PSYCHOLOGY)

A fantasy is a situation imagined by an individual or group, which does not correspond with reality but expresses certain desires or aims of its creator. Fantasies typically involve situations which are impossible (such as the existence of magic powers) or highly unlikely (such as world peace). Fantasies can also be sexual in nature.

Psychoanalysis

In the theory of psychoanalysis, phantasy is used to describe unconscious desires, fears, drives etc. Sigmund Freud used the German word 'Phantasie', which could be translated as 'fantasy', but the meaning is clearly not the same as the everyday meaning and is usually printed as 'phantasy'. This should be strongly contrasted with delusion.

12

Self Psychology

The self is a key construct in several schools of psychology. Ideas are different for many theorist and in fields of study, but in general the self refers to the conscious reflective personality of a person. If we develop a higher thinking for ourselves the beliefs and values between the real and ideal self will expand and therefore we can make more developed morals and reasons, when we know who we are. The study of the self involves significant methodological problems, especially concerning consciousness. Some of these are taken up in philosophy of mind and metaphysics. A psychological school of thought focused on the self was originally proposed by Heinz Kohut (1913-1981).

IDENTITY CRISIS

Erikson, the psychologist who coined the term identity crisis, believes that the identity crisis is the most important conflict human beings encounter when they go through eight developmental stages in life. The identity is "a subjective sense as well as an observable quality of personal sameness and continuity, paired with some belief in the sameness and continuity of some shared world image. As a quality of unself-conscious living, this can be gloriously obvious in a young person who has found himself as he has found his communality. In him we see emerge a unique unification of what is irreversibly given--that is, body type and temperament, giftedness and vulnerability, infantile models and acquired ideals--with the open choices provided in available roles, occupational possibilities, values offered, mentors met, friendships made, and first sexual encounters." (Erikson, 1970.)

According to Erikson's stages, the onset of the identity crisis is in the teenage years, and only individuals who succeed in resolving the crisis will be ready to face future challenges in life. But the identity crisis may well be recurring, as the changing world demands us to constantly redefine ourselves. Erikson suggested that people experience an identity crisis when they lose "a sense of personal sameness and historical continuity". Given today's rapid development in technology, global economy, dynamics in local and world politics, identity crises are expected to be more common now than 30 years ago, when Erikson formed his theory. If you find yourself (again) in an identity crisis, you can look at seven areas of difficulty in which to work towards a resolution.

Time Perspective: Can you distinguish immediate gratification from long-term goals? Have you learnt to balance between jumping at opportunities as soon as they are presented to you and working steadily and patiently towards your long-term goal?

Self-Certainty: Do you feel consistent in your self-image and the image you present to others?

Role Experimentation: Have you tried different roles in search of the one that feels right to you?

Anticipation of Achievement: Do you believe that you will be successful in what you choose to do--whether your role is at the work front or home front?

Sexual Identity: Do you feel comfortable being a male or a female, and dealing with others as such?

Leadership Polarization: Are you able to become both a leader and a follower, whichever is called for in a given situation?

Idealogical: Have you found a set of basic social, philosophical, or religious values that your outlook on life can be based upon?

IMITATION

Imitation is an advanced behavior whereby an individual observes and replicates another's. The word can be applied in many contexts, ranging from animal training to international politics.

In anthropology, diffusion theories explain why cultures imitate the ideas or practices of other cultures. Some theories hold that all cultures imitate ideas from one or a few original cultures, the Adam of the Bible, or several cultural circles that overlap. Evolutionary diffusion theory holds that cultures are influenced by one another, but that similar ideas can be developed in isolation.

In music, it refers to the repetition of a phrase played on one instrument by another instrument.

In mid-20th century, social scientists began to study how and why people imitate ideas. Everett Rogers pioneered diffusion of innovations studies, using research to prove factors in adoption and profiles of adopters of ideas.

Recent work in neuroscience has begun to reveal the mechanisms of imitation in the human brain. It seems that there is a system of mirror neurons which are active both when you see another person act and when you do the same thing yourself, and these may allow humans to learn by imitation.

It has been argued by Susan Blackmore in The Meme Machine, that imitation is what makes humans unique among animals. Imitation might have been selected as fit by evolution because those who were good at it had a wider arsenal of learned cultural behavior at their disposal, such as tool making or even language.

INTROSPECTION

Introspection is contemplation on one's self, as opposed to extrospection, the observation of things external to one's self. Introspection may be used synonymously with self-reflection and used in a similar way. Introspection is like the activity described by Plato when he asked, "...why should we not calmly and patiently review our own thoughts, and thoroughly examine and see what these appearances in us really are?" (Theaetetus, 155)

Cognitive psychology accepts the use of the scientific method, but rejects introspection as a valid method of investigation. It should be noted that Herbert Simon and Allen Newell identified the 'thinking-aloud' protocol, in which investigators view a subject engaged in introspection, and who speaks his thoughts aloud, thus allowing study of his introspection.

Introspection was once an acceptable means of gaining insight into psychological phenomena. Introspection was used by German physiologist Wilhelm Wundt in the experimental psychology laboratory he had founded in Leipzig in 1879. Wundt believed that by using introspection in his experiments he would gather information into how the subject's minds were working, thus he wanted to examine the mind into its basic elements. Wundt did not invent this way of looking into an individual's mind through their experiences; rather, it can date to Socrates. Wundt's distinctive contribution was to take this method into the experimental arena and thus into the newly formed field of psychology.

SELF-AWARENESS

Self-awareness is the explicit understanding that one exists. Furthermore, it includes the concept that one exists as an individual, separate from other people, with private thoughts. It may also include the understanding that other people are similarly self-aware. Self-awareness remains a critical mystery in philosophy, psychology, biology, and artificial intelligence. Self-awareness is a unique type of consciousness, in that it is not always present.

Self-consciousness is credited with the development of identity. In an epistemological sense, self-consciousness is a personal understanding of the very core of one's own identity. It is during periods of self-consciousness that people come the closest to knowing themselves objectively. Jean Paul Sartre describes self-consciousness as being "non-positional", in that it is not from any location in particular.

Self-consciousness plays a large role in behaviour as it is common to act differently when people "lose one's self in a crowd". It is the basis for human traits, such as accountability and conscientiousness. It also plays a large role in religion, and existentialism. Self-consciousness affects people in varying degrees, as some people self-monitor (or scrutinize) themselves more than others. Different cultures vary in the importance they place on self-consciousness.

SELF AWARENESS AND SELF CONSCIOUSNESS

Self-awareness

Human self-awareness leads us to recognize three core paradoxes or absurd features of the human condition:

- The human imagination has no physical boundaries, but our bodies do. In our minds, we can instantly travel to the ends of the universe, the center of the earth, even the center of the sun. We can use our mental microscope to visualize germs, viruses, atoms, quarks. As soon as we detect something with any instrument, we can make images of it in our minds. We travel effortlessly in our thoughts. The boundless production of fiction literature is evidence of the creative powers of the human imagination. Yet physically we are bound to one specific, small planet, and due to the speed limit of the universe (speed of light), it appears that we are bound to a small neighborhood around this planet for the foreseeable future. This paradox is the physical frustration of the human condition.
- Human spirits can motivate the noblest and holiest thoughts, the most altruistic actions, the most beneficial generosities. But they can also produce the most horrible cruelties and violence against countless people, including suicide of the perpetrators. Our will effortlessly moves our thoughts one way and then another, untamed by moral law or conscience. Leaders can sway whole populations to do things--benevolent or malevolent--that individuals would never, on their own, have contemplated. How can these two extremes coexist in the same individual? We don't observe such extremes in other animals. They are exclusive to the human condition.
- Human actions and our very lives are motivated by hope--that we can make a difference, that we can learn and grow and build and make things better. Yet

physically speaking we know that we are mortal, we are made of dust, and we will return to dust. Despite this realization, hope springs eternal. Without hope, as Albert Camus said, the only serious philosophical question is why we should not commit suicide. Hope gets us up in the morning, and drives us forward every day. By extension, we hope for eternal life beyond the grave--God, heaven, paradise--because otherwise our existential situation has no meaning. These aspirations--for hope, meaning, significance, purpose, identity, peace, happiness, beauty, love--are all aspects of human spirituality.

These paradoxes present universal, inescapable questions about life. Whenever any people lift their thoughts from daily routines, they may ponder these questions. Attempts to explain or resolve these paradoxes are the domain of religion and philosophy. The human condition is the central subject of all literature, drama and art.

Consciousness and self-awareness are poorly understood. Self-awareness is a unique type of consciousness in that it is not always present, and is not sought after. Meditation or repetitive tasks, as well as some schools of thought in art theory and existentialism seek to reduce self-consciousness, and even self-awareness, at least temporarily.

The ability to self-analyze (or scrutinize) is widely believed among psychologists not to develop until mid-childhood, and arguably is present in only a few species of animals. Tests that are usually considered as representative of self-consciousness include applying a bright dot to a subjects forehead, and then placing them in front of a mirror-if they reach for their own forehead, it appears they may realize their own existence in a self-aware sense. Great apes, dolphins and elephants can pass this test. However, others criticize this test as testing only the understanding of the duplicability of image, and not especially meaningful as a way of gauging self-consciousness.

Suffering in the Zen Buddhist identifies negative results from attaching firmly to the narrow conception of a self that is an unchanging entity. For example: yesterday's self was healthy and happy but today's self is ill and lamenting the loss of health in addition to suffering with the pain of ill health. The philosopher Ludwig Feuerbach's theory of religion was based upon projection deriving from a Hegelian sense of self consciousness.

Self Consciousness

The term self conscious has a different meaning in colloquial use, namely a person who is worried or apprehensive how they may appear to others. It is widely believed that this trait is most present during the teenage years.

When one is feeling self-conscious, one can feel too aware of even the smallest of one's own actions. Such awareness can impair one's ability to perform complex actions. For example, a piano player may "choke", lose confidence, and even lose the ability to perform when they notice the audience. As self-consciousness fades one may regain the ability to focus.

A person who is especially prone to self-awareness may be labeled shy or introverted. Work has been done in the area of flow psychology-attempts to escape self-consciousness. Buddhism, as well as some schools of thought in art theory and existentialism also aim to reduce self-consciousness, at least temporarily.

Unlike self-awareness, self-consciousness has connotations of being unpleasant, and is often linked to self-esteem. Self-consciousness is credited with the development of identity, because it is during periods of self-consciousness that people come the closest to knowing themselves objectively. Self-consciousness plays a large role in behavior, as it is common to act differently when people "lose themselves in a crowd". Self-consciousness affects people in varying degrees, as some people are in constant self-monitoring, while others are completely oblivious about themselves.

The Basis of Personal Identity

John Locke's chapter XXVII "On Identity and Diversity" in An Essay Concerning Human Understanding (1689) has been said to be one of the first modern conceptualizations of consciousness as the repeated self-identification of oneself, through which moral responsibility could be attributed to the subject-and therefore punishment and guiltiness justified, as would critics such as Nietzsche point out. John Locke does not use the terms self-awareness or self-consciousness though.

According to Locke, personal identity (the self) "depends on consciousness, not on substance" nor on the soul. We are the same person to the extent that we are conscious of our past and future thoughts and actions in the same way as we are conscious of our present thoughts and actions. If consciousness is this "thought" which doubles all thoughts, then personal identity is only founded on the repeated act of consciousness: "This may show us wherein personal identity consists: not in the identity of substance, but... in the identity of consciousness".

For example, one may claim to be a reincarnation of Plato, therefore having the same soul. However, one would be the same person as Plato only if one had the same consciousness of Plato's thoughts and actions that he himself did. Therefore, self-identity is not based on the soul. One soul may have various personalities. Self-identity is not founded either on the body or the substance, argues Locke, as the substance may change while the person remains the same: "animal identity is preserved in identity of life, and not of substance", as the body of the animal grows and changes during its life. Take for example a prince's soul which enters the body of a cobbler: to all exterior eyes, the cobbler would remain a cobbler.

But to the prince himself, the cobbler would be himself, as he would be conscious of the prince's thoughts and acts, and not of the cobbler's life. A prince's consciousness in a cobbler body: thus the cobbler is, in fact, a prince. But this interesting border-case leads to this problematic thought that since personal identity is based on consciousness, and that only oneself can be aware of his consciousness, exterior human judges may never know if they really are judging-and punishing-the same person, or simply the same body. In other words, Locke argues that you may be judged only for the acts of your body, as this is what is apparent to all but God; however, you are in truth only responsible for the acts for which you are conscious. This forms the basis of the insanity defense: one can't be held accountable for acts from which one was unconscious-and therefore leads to interesting philosophical questions: "personal identity consists [not in the identity of substance] but in the identity of consciousness, wherein if Socrates and the present mayor of Queenborough agree, they are the same person: if the same Socrates waking and sleeping do not partake of the same consciousness, Socrates waking and sleeping is not the same person. And to punish Socrates waking for what sleeping Socrates thought, and waking Socrates was never conscious of, would be no more right, than to punish one twin for what his brother-twin did, whereof he knew nothing, because their outsides were so like, that they could not be distinguished; for such twins have been seen."

Or again:

"PERSON, as I take it, is the name for this self. Wherever a man finds what he calls himself, there, I think, another may say is the same person. It is a forensic term, appropriating actions and their merit; and so belong only to intelligent agents, capable of a law, and happiness, and misery. This personality extends itself beyond present existence to what is past, only by consciousness,--whereby it becomes concerned and accountable; owns and imputes to itself past actions, just upon the same ground and for the same reason as it does the present. All which is founded in a concern for happiness, the unavoidable concomitant of consciousness; that which is conscious of pleasure and pain, desiring that that self that is conscious should be happy. And therefore whatever past actions it cannot reconcile or APPROPRIATE to that present self by consciousness, it can be no more concerned in it than if they had never been done: and to receive pleasure or pain, i.e. reward or punishment, on the account of any such action, is all one as to be made happy or miserable in its first being, without any demerit at all. For, supposing a MAN punished now for what he had done in another life, whereof he could be made to have no consciousness at all, what difference is there between that punishment and being CREATED miserable? And therefore, conformable to this, the apostle tells us, that, at the great day, when every one shall 'receive according to his doings, the secrets of all hearts shall be laid open.' The sentence shall be justified by the consciousness all person shall have, that THEY THEMSELVES, in what bodies soever they appear, or what substances soever that consciousness adheres to, are the SAME that committed those actions, and deserve that punishment for them."

Henceforth, Locke's conception of personal identity founds it not on the substance or the body, but in the "same continued consciousness", which is also distinct from the soul since the soul may have no consciousness of itself (as in reincarnation). He creates a third term between the soul and the body-and Locke's thought may certainly be meditated by those who, following a scientist ideology, would identify too quickly the brain to consciousness. For the brain, as the body and as any substance, may change, while consciousness remains the same. Therefore personal identity is not in the brain, but in consciousness. However, Locke's theory also reveals his debt to theology and to Apocalyptic "great day", which by advance excuse any failings of human justice and therefore humanity's miserable state.

Physiological Location for Self-awareness

A functional magnetic resonance imaging (fMRI) experiment by Ilan Goldberg et. al. (April 19, 2006) in Neuron (vol 50, p 329) has demonstrated the functional separation of sensory processing and self-awareness. Self-awareness appears to be processed in the superior frontal gyrus.

Self-awareness in Theater

In theater, self-awareness refers to a fictional character who is depicted as breaking character, perhaps by breaking the fourth wall. In some plays, such as Six Characters in Search of an Author and Kiss Kiss Bang Bang, this may be associated with inviting the audience to continue their suspension of disbelief; that is to say, to believe that the characters depicted by the cast are "really" self-aware fictional characters who know that they are fictional characters, but can do nothing about it.

Theatre also concerns itself with other awareness besides self-awareness. There is a possible fractal correlation between the experience of the theater audience and individual self-awarenesss. As actors and audiences must not 'break' the fourth wall in order to maintain context, so individuals must not be aware of the artificial, or the constructed perception of his or her reality. This suggests that self-awareness is an artificial continuum just as theater is. Theatrical efforts such as Six Characters in Search of an Author or say, The Wonderful Wizard of Oz, construct yet another layer of the fourth wall, but they do not destroy the primary illusion. Refer to Erving Goffman's Frame Analysis: An Essay on the Organization of Experience.

CONFIDENCE

Confidence is trust or faith that a person or thing is capable. Self-confidence is having confidence in oneself. Overconfidence is having unmerited confidence-believing something or someone is capable when they are not. People may have confidence in other people or forces beyond their control. For instance, one might have confidence in the police to protect them, or may have confidence that a sports team will win a game. Faith and Trust are synonyms of confidence when used in this sense.

Usually when someone is referred to as 'confident' they are referring to self-confidence. Self-confidence is faith in one's own abilities. People with high self-confidence typically have little fear of the unknown, are able to stand up for what they believe in, and have the courage to risk embarrassment (for instance, by giving a presentation to a large group of people). One who is self-confident is not necessarily loud, brash, or reckless.

Losing Confidence

Losing confidence is no longer trusting in the ability to perform. It may be reasonable as the result of past failure to perform, or unreasonable, because one "just has a feeling" about something or is having doubt.

Choking refers to losing confidence, especially self-confidence, just at the moment when it is needed most and doing poorly as a result e.g. in sports. This is found as a common plot device in literature and film, and is usually devised to result in a total alteration of a character's life.

Confidence in the face of danger is also known as bravery.

The states of confidence could thus be described as:

Overconfident: In the absence of anxiety a person could become reckless due to overconfidence. This state can be identified by the cognitive thought process of everything appearing rosy. An overconfident person doesn't see the need to consider all possible outcome and is sure the outcome will be what he/she has perceived.

Unconfident: An unconfident person on the other hand gets paralyzed due to anxiety. The person tends to have lots of self-doubts, becomes very critical about himself / herself and have low opinion of self (low self esteem). This causes a vicious cycle where the anxiety cripples the person into inaction, and he / she continues to berate herself thus creating more anxiety.

Confidence: When the anxiety is at an optimum level, you are at your best. You know that negative outcomes are possible, but rather than exaggerating or minimizing it, you give it the due attention necessary (what can I do if this happens...). So perhaps a better definition of confidence is the state of balanced perceptions and preparation.

Self-consciousness

Self-consciousness is an acute sense of self-awareness. It is a preoccupation with oneself, rather than the philosophical state of self-awareness, which is the awareness that one exists as an individual being. An unpleasant feeling of self-consciousness occurs when we realize that we are being watched or observed, the feeling that "everyone is looking" at us. Some people are habitually more self-conscious than others. Feelings of self-consciousness are sometimes associated with shyness or paranoia.

Impairment

When feeling self-conscious, one becomes aware of even the smallest of one's own actions. Such awareness can impair one's ability to perform complex actions. For example, a piano player may "choke", lose confidence, and even lose the ability to perform at the moment they notice the audience. This is a function of the psychological phenomenon of social facilitation. As self-consciousness fades one may regain the ability to "lose one's self".

Adolescence is believed to be a time of heightened self-consciousness. A person with a chronic tendency toward self-consciousness may be shy or introverted.

Psychology

Unlike self-awareness, self-consciousness can be a problem at times. It is often associated with shyness and embarrassment, and can affect self-esteem. In a positive context, self-consciousness may affect the development of identity, because it is during periods of high self-consciousness that people come the closest to knowing themselves objectively. Self-consciousness affects people in varying degrees, as some people are constantly self-monitoring or self-involved, while others are completely oblivious about themselves.

Psychologists frequently distinguish between two kinds of self-consciousness, private and public. Private self-consciousness is a tendency to introspect and examine one's inner self and feelings. Public self-consciousness is an awareness of the self as it is viewed by others. This kind of self-consciousness can result in self-monitoring and social anxiety. Both private and public self-consciousness are viewed as personality traits that are relatively stable over time, but they are not correlated. Just because an individual is high on one dimension doesn't mean that he or she is high on the other.

Different levels of self-consciousness affect behavior, as it is common for people to act differently when they "lose themselves in a crowd". Being in a crowd, being in a dark room, or wearing a disguise creates anonymity and temporarily decrease self-consciousness. This can lead to uninhibited, sometimes destructive behavior. Placing an individual in front of a mirror can temporarily increase self-consciousness.

SELF-CONCEPT

Self-concept or self-identity is the mental and conceptual understanding and persistent regard that sentient beings hold for their own existence. In other words, it is the sum total of a being's knowledge and understanding of his or her self. The self-concept is different from self-consciousness, which is an awareness or preoccupation with one's self. Components of the self-concept include physical, psychological, and social attributes, which can be influenced by the individual's attitudes, habits, beliefs and ideas. These components and attributes can be condensed to the general concepts of self-image and the self-esteem.

History

A milestone in human reflection about the non-physical inner self came in 1644, when René Descarıes wrote Principles of Philosophy. Descartes proposed that doubt was a principal tool of disciplined inquiry, yet he could not doubt that he doubted. He reasoned that if he doubted, he was thinking, and therefore he must exist. Thus existence depended upon perception.

A second milestone in the development of self-concept theory was the writing of Sigmund Freud (1900) who gave us new understanding of the importance of internal mental processes. While Freud and many of his followers hesitated to make self-concept a primary psychological unit in their theories, Freud's daughter Anna (1946) gave central importance to ego development and self-interpretation.

Self-concept theory has always had a strong influence on the emerging profession of counseling. Prescott Lecky (1945) contributed the notion that self-consistency is a primary motivating force in human behavior. Raimy (1948) introduced measures of self-concept in counseling interviews and argued that psychotherapy is basically a process of altering the ways that individuals see themselves.

By far the most influential and eloquent voice in self-concept theory was that of Carl Rogers (1947) who introduced an entire system of helping built around the importance of the self. In Rogers' view, the self is the central ingredient in human personality and personal adjustment.

Rogers described the self as a social product, developing out of interpersonal relationships and striving for consistency. He maintained that there is a basic human need for positive regard both from others and from oneself. He also believed that in every person there is a tendency towards self-actualization and development so long as this is permitted and encouraged by an inviting environment (Purkey & Schmidt, 1987).

While most self-concept theorists continued to write and conduct research during the 1970s and 1980s, general interest in self-concept declined. In a recent article explaining the likely causes for the decline of "humanistic" education, Patterson (1987) presents reasons for the decline of interest in self-concept as well. He offers four likely causes:

- A cornucopia of contrived games, gimmicks, and techniques that were introduced and controlled by unprepared professionals.
- A national mood of "back to basics" in education prevailed where concern for the emotional needs of students was viewed as inimical to academic excellence.
- Poor judgment by counselors and teachers in selecting suitable materials for values clarification programs resulted in public opposition to any attempt to introduce values in school.
- Strong opposition by those who objected to any consideration of personal development of students because they believed it to be secular humanism and, therefore, an effort to undermine religion.

Fortunately, there is a new awareness on the part of both the public and professionals that self-concept cannot be ignored if we are to successfully address such nagging problems as drug and alcohol abuse, drop-out rates, dysfunctional families, and other concerns. In addition to this growing awareness, new ways are being developed to strengthen self-concepts.

For example, research by cognitive theorists (McAdam, 1986; Ryan, Short & Weed, 1986) are demonstrating that negative self-talk leads to irrational thinking regarding oneself and the world.

Basic Assumptions

Many of the successes and failures that people experience in many areas of life are closely related to the ways that they have learned to view themselves and their relationships with others. It is also becoming clear that self-concept has at least three major qualities of interest to counselors: (1) it is learned, (2) it is organized, and (3) it is dynamic. Each of these qualities, with corollaries, follow.

Self-concept is learned. As far as we know, no one is born with a self-concept. It gradually emerges in the early months of life and is shaped and reshaped through repeated perceived experiences, particularly with significant others. The fact that self-concept is learned has some important implications:

- Because self-concept does not appear to be instinctive, but is a social product developed through experience, it possesses relatively boundless potential for development and actualization.
- Because of previous experiences and present perceptions, individuals may perceive themselves in ways different from the ways others see them.
- Individuals perceive different aspects of themselves at different times with varying degrees of clarity. Therefore, inner focusing is a valuable tool for counseling.
- Any experience which is inconsistent with one's self-concept may be perceived as a threat, and the more of these experiences there are, the more rigidly self-concept is organized to maintain and protect itself. When a person is unable to get rid of perceived inconsistencies, emotional problems arise.
- Faulty thinking patterns, such as dichotomous reasoning (dividing everything in terms of opposites or extremes) or overgeneralizing (making sweeping conclusions based on little information) create negative interpretations of oneself.

Self-concept is organized. Most researchers agree that self-concept has a generally stable quality that is characterized by orderliness and harmony.

Each person maintains countless perceptions regarding one's personal existence, and each perception is orchestrated with all the others. It is this generally stable and organized quality of self-concept that gives consistency to the personality. This organized quality of self-concept has corollaries.

- Self-concept requires consistency, stability, and tends to resist change. If self-concept changed readily, the individual would lack a consistent and dependable personality.
- The more central a particular belief is to one's self-concept, the more resistant one is to changing that belief.
- At the heart of self-concept is the self-as-doer, the "I," which is distinct from the self-as-object, the various "me's." This allows the person to reflect on past events, analyze present perceptions, and shape future experiences.
- Basic perceptions of oneself are quite stable, so change takes time.
- Perceived success and failure affect self-concept. Failure in a highly regarded area lowers evaluations in all other areas as well. Success in a prized area raises evaluations in other seemingly unrelated areas.

Self-concept is dynamic. To understand the active nature of self-concept, it helps to imagine it as a gyrocompass: a continuously active system that dependably points to the "true north" of a person's perceived existence. This guidance system not only shapes the ways a person views oneself, others, and the world, but it also serves to direct action and enables each person to take a consistent "stance" in life. Rather than viewing self-concept as the cause of behavior, it is better understood as the gyrocompass of human personality, providing consistency in personality and direction for behavior. The dynamic quality of self-concept also carries corollaries.

- The world and the things in it are not just perceived; they are perceived in relation to one's self-concept.
- Self-concept development is a continuous process. In the healthy personality there is constant assimilation of new ideas and expulsion of old ideas throughout life.
- Individuals strive to behave in ways that are in keeping with their self-concepts, no matter how helpful or hurtful to oneself or others.

- Self-concept usually takes precedence over the physical body. Individuals will often sacrifice physical comfort and safety for emotional satisfaction.
- Self-concept continuously guards itself against loss of self-esteem, for it is this loss that produces feelings of anxiety.
- If self-concept must constantly defend itself from assault, growth opportunities are limited.

Summary

This brief overview of self-concept theory has focused on describing the ways people organize and interpret their inner world of personal existence. The beginnings of self-concept theory and its recent history have been discussed. Three major qualities of self-concept--that it is: (1) learned, (2) organized, and (3) dynamic--have been presented. Individuals have within themselves relatively boundless potential for developing a positive and realistic self-concept. This potential can be realized by people, places, policies, programs, and processes that are intentionally designed to invite the realization of this potential.

Bibliography

Adas, Michael: *Machines as the Measure of Men: Science, Technology, and Ideologies of Western Dominance*, Ithaca, Cornell University Press, 1989.

Aggarwal, J.C.: *National Policy on Education: Agenda for India 2001*, New Delhi, Concept, 1989.

Ali, S. Mahmud: *The Fearful State: Power, People, and Internal War in South Asia*, London, Zed, 1993.

Atwood, G. and Stolorow R.: *Structures of Subjectivity*, Hillsdale, The Analytic Press, 1984.

Awasthi, Dipesh: *Regional Patterns of Industrial Growth in India*, New Delhi, Concept, 1991.

Bacal, H. & Newman, K.: *Theories of Object Relations: Bridges to Self Psychology*, New York, Columbia University Press, 1990.

Bag, A.K.: *Science and Civilization in India*, New Delhi, Navrang, 1985.

Banu, Zenab: *Politics of Communalism*, Bombay, Popular Prakashan, 1989.

Bardhan, Pranab K.: *The Political Economy of Development in India*, Delhi, Oxford University Press, 1984.

Basch, M. F.: *Doing Psychotherapy*, New York, Basic Books, 1980.

Cohn, Bernard S.: *India: The Social Anthropology of a Civilization*, New Jersey: Prentice-Hall, 1971.

Datta, Prabhat Kumar: *Regionalism of Indian Politics*, New Delhi, Sterling, 1993.

Detrick, D. & Detrick, S.: *Self Psychology: Comparisons and Contrasts*, Hillsdale, Analytic Press, 1989.

Dewald, P.A.: *Psychotherapy: A Dynamic Approach*, New York, Basic Books, 1971.

Dhar, Pannanlal N.: *Preventive Detention under Indian Constitution*, New Delhi, Deep and Deep, 1986.

Dixit, Prabla: *Communalism: A Struggle for Power*, New Delhi, Orient Longman, 1981.

Elson, M.: *Self Psychology in Clinical Social Work*, New York, W.W. Norton & Company, 1986.

Freud, S.: *Project for a Scientific Psychology*, London,Hogarth Press, 1966.

Gedo, J. E. & Goldberg, A.: *Models of the Mind: A Psychoanalytic Theory*, Chicago, University of Chicago Press, 1973.

Goldberg, A.: *The Psychology of the Self: A Casebook*, New York, International Universities Press, 1978.

Greenson, R. R.: *The Technique and Practice of Psychoanalysis*, New York, International Universities Press, 1967.

Kohli, Shanta: *Family Planning in India: A Descriptive Analysis*, New Delhi, Indian Institute of Public Administration, 1977.

Kohut, H.: *Self Psychology and the Humanities*, New York, W. W. Norton, 1985.

Lee, R.& Martin, C.: *Psychotherapy after Kohut: A Textbook of Psychology*, Hillsdale, The Analytic Press, 1991.

Lichtenberg. J. D.: *Psychoanalysis and Motivation*, Hillsdale, The Analytic Press, 1989.

Magid, B.: *Freud's Case Studies: Self-Psychological Perspectives*, Hillsdale, Analytic Press, 1993.

Mahler, M. S., Pine, F.: *The Psychological Birth of the Human Infant*, New York, Basic Books, 1975.

Mathur, V.S.: *Studies in Indian Education*, New Delhi, Arya Book Depot, 1968.

Parens, H.: *The Development of Aggression in Early Childhood*, New York, Aronson, 1979.

Sankaia, H.D.: *Aspects of Indian History and Archeology*, Delhi: B.R. Publishing, 1977.

Schafer, R.: *A New Language for Psychoanalysis*, New Haven, Yale University Press, 1976.

Shapiro, S.: *Talking with Patients: A Self Psychological View of Creative Intuition and Analytic Discipline*, Northvale, Jason Aronson Inc, 1995.

Spence, D.: *Narrative Truth and Historical Truth*, New York, Norton, 1982.

Stepansky, P. E. & Goldberg, A.: *Kohut's Legacy: Contributions to Self Psychology*, Hillsdale, Analytic Press, 1984.

Stolorow, R.: *Psychoanalytic Treatment: An Intersubjective Approach*, Hillsdale, The Analytic Press, 1987.

Strozier, Charles B.: *Heinz Kohut, Self Psychology and The Humanities: Reflections on a New Psychoanalytic Approach*, New York, Norton Publishing Company, 1985.

Suth, R.: *Progress in Self Psychology*, Hillsdale, Analytic Press, 1996.

Terman, D.: *Object Love and the Psychology of the Self, New York*, International Universities Press, 1980.

Tolpin, P.: *The Borderline Personality: Its Makeup and Analyzability*, New York, International Universities Press, 1980.

Ulman, R., & Brothers, D.: *The Shattered Self: A Psychoanalytic Study of Trauma*, Hillsdale, Analytic Press, 1988.

Wolf, E.S.: *Selfobject Experiences: Development, Psychopathology, Treatment. Presented to 24th Annual Margaret Mahler Symposium on Child Development*, Northvale, Jason Aronson, 1994.

Wolf, E. S.: *Treating the Self*, New York, The Guilford Press, 1988.

Index

D

E

F

M

N

O

P

□□□